Claus Kuhnel

BASCOM
Programming of Microcontrollers with Ease

An Introduction
by Program Examples

BASCOM Programming of Microcontrollers with Ease:
An Introduction by Program Examples

Copyright © 2001 Claus Kuhnel
All rights reserved. No part of this work may be reproduced in any
form except by written permission of the author.
All rights of translation reserved.

Publisher and author assume no responsibility for any errors that
may arise from the use of devices and software described in this
book.

Universal Publishers/uPUBLISH.com
USA • 2001

ISBN: 1-58112-671-9

www.upublish.com/books/kuhnel.htm

Preface

The microcontroller market knows some well introduced 8-bit microcontroller families like Intel's 8051 with its many derivatives from different manufacturers, Motorola's 6805 and 68HC11, Microchip's PICmicros and Atmel's AVR.

The 8051 microcontroller family has been well-known over many years. The development of new derivatives is not finished yet. From time to time new powerful derivatives are announced.

You will find derivatives from Philips, Dallas, Analog Devices and Cygnal and others with the known 8051 core but enhanced clock and peripherals. For example, complete analog-to-digital and digital-to-analog subsystems were integrated in some chips.

Atmel developed the AVR microcontroller family which is well suited for high-level language programming and in-system programming.

For all those microcontrollers there is development software ranging from simple assemblers for DOS to integrated development environments for Windows95/98/NT on the market.

Apart from programming environments as they are offered, for example, by KEIL, IAR or E-LAB Computer for professional applications, also the more economical and nonetheless sufficiently equipped development environments can maintain ground.

BASCOM-8051 and BASCOM-AVR are development environments built around a powerful BASIC compiler which is suited for project handling and program development for the 8051 family and its derivatives as well as for the AVR microcontrollers from Atmel.

The programming of microcontrollers using BASCOM-8051 (version 2.0.4.0) and BASCOM-AVR (version 1.11.3.0) will be described in this book.

Some applications help understand the usage of BASCOM-8051 and BASCOM-AVR.

Acknowledgement

I should like to thank the following:

- in the first place, Mark Alberts of MCS Electronics, who developed the BASCOM programming environment at an outstanding price-performance ratio,
- Atmel for the development of the AVR RISC microcontrollers which introduced new capabilities into the microcontroller families,
- Christer Johansson of High Tech Horizon, who supports safe communication of microcontrollers and PC by the development and free distribution of the S.N.A.P. protocol and the necessary tools effectively and
- Lars Wictorsson of LAWICEL for the development of the CANDIPs, microcontroller modules with CAN interface.

Contents

1 Supported Microcontrollers 9
  1.1 8051 Family 9
  1.2 AVR Family 11
2 BASCOM 23
  2.1 BASCOM Demos 23
  2.2 BASCOM Commercial Versions 25
  2.3 Update of BASCOM Commercial Versions 25
  2.4 BASCOM Projects 27
    2.4.1 Working on Projects 27
    2.4.2 BASCOM Options 28
  2.5 BASCOM Tools 37
    2.5.1 Simulation 37
    2.5.2 Terminal Emulator 40
    2.5.3 LCD Designer 42
    2.5.4 Library Manager 46
    2.5.5 Programming Devices 50
  2.6 Hardware for AVR RISC Microcontroller 55
    2.6.1 DT006 AVR Development Board 55
    2.6.2 AVR-ALPHA with AT90S2313 56
  2.7 Instead of "Hello World" 57
    2.7.1 AVR 57
    2.7.2 8051 58
    2.7.3 Things in Common 59
    2.7.4 Simulation 64
  2.8 BASCOM Help System 67
3 Some BASCOM Internals 69
  3.1 Building new instructions 69

3.2 Parameters for Subroutines in BASCOM-AVR 71
3.3 BASIC & Assembler 73
    3.3.1 AVR 74
    3.3.2 8051 75
4 Applications 77
  4.1 Programmable Logic 77
  4.2 Timer and Counter 81
    4.2.1 AVR 81
    4.2.2 8051 104
  4.3 LED Control 107
    4.3.1 Single LED 107
    4.3.2 Seven-Segment Displays 108
    4.3.3 Dot-Matrix Displays 114
  4.4 LCD Control 119
    4.4.1 Direct Control 119
    4.4.2 LCD with Serial Interface 122
  4.5 Connecting Keys and Keyboards 128
    4.5.1 Single Keys 129
    4.5.2 Matrix Keypad 132
    4.5.3 PC-AT Keyboard 136
  4.6 Data Input by IR Remote Control 140
  4.7 Asynchronous Serial Communication 143
  4.8 1-WIRE Interface 151
  4.9 SPI Interface 161
  4.10 I^2C Bus 167
  4.11 Scalable Network Protocol S.N.A.P. 173
    4.11.1 S.N.A.P. Features 174
    4.11.2 Description of S.N.A.P. Protocol 175
    4.11.3 S.N.A.P. Monitor 179
    4.11.4 Digital I/O 183

- 4.12 CANDIP - Interface to CAN 197
- 4.13 Random Numbers 209
- 4.14 Moving Average 214
- 5 Appendix 219
  - 5.1 Decimal-Hex-ASCII Converter 219
  - 5.2 DT006 Circuit Diagram 220
  - 5.3 Characters in Seven-Segment Display 222
  - 5.4 BASIC Stamp II 223
  - 5.5 Literature 224
  - 5.6 Links 225
- 6 Index 231

1 Supported Microcontrollers

BASCOM is an Integrated Development Environment (IDE) that supports the 8051 family of microcontrollers and some derivatives as well as Atmel's AVR microcontrollers. Two products are available for the various microcontrollers - BASCOM-8051 and BASCOM-AVR.

In a microcontroller project one needs to know the hardware base, i.e. the microcontroller with internal and connected peripherals, and the software used, i.e. IDE handling, programming and debugging.

In this first chapter, let's have a look at the supported microcontrollers. A general overview will be given only; the various parts are documented by the manufacturers in more detail. You may also search the web for more information and documentation on all the microcontrollers dealt with here.

1.1 8051 Family

The 8051 is an accumulator-based microcontroller featuring 255 instructions. A basic instruction cycle takes 12 clocks; however, some manufacturers redesigned the instruction-execution circuitry to reduce the instruction cycle.

The CPU has four banks of eight 8-bit registers in on-chip RAM for context switching. These registers reside within the 8051's lower 128 bytes of RAM along with a bit-operation area and scratchpad RAM. These lower bytes can be addressed directly or indirectly by using an 8-bit value. The upper 128 bytes of on-chip data RAM encompass two overlapping address spaces. One space is for directly addressed special-function registers (SFRs); the other space is for indirectly addressed RAM or stack. The SFRs define peripheral operations and configurations. The 8051 also has 16 bit-addressable bytes of on-chip RAM for flags or variables.

Without external circuitry, the maximum address range of all 8051 processors is 64 Kbytes of program memory and 64 Kbytes of data memory. External means can be made use of to increase this address space.

Register indirection uses an 8-bit register for an on-chip RAM address; an off-chip address requires an 8- or 16-bit data-pointer register (DPTR). The original 8051 has only one DPTR. Derivatives from Atmel, Dallas, and Philips have two DPTRs. Siemens microcontrollers have eight DPTRs. The 8051 microcontroller has bidirectional and individually addressable I/O lines.

The 8051 performs extensive bit manipulation via instructions, such as set, clear, complement, and jump on bit set or jump on bit clear, only for a 16-byte area of RAM and some SFRs. It can also handle AND or OR bits with a carry bit. The Dallas versions have variable-length move-external-data instructions. Math functions include add, subtract, increment, decrement, multiply, divide, complement, rotate, and swap nibbles. Some of the Siemens devices have a hardware multiplier/divider for 16-bit multiply and 32-bit divide. Figure 1 shows the block diagram of an 8051 [1].

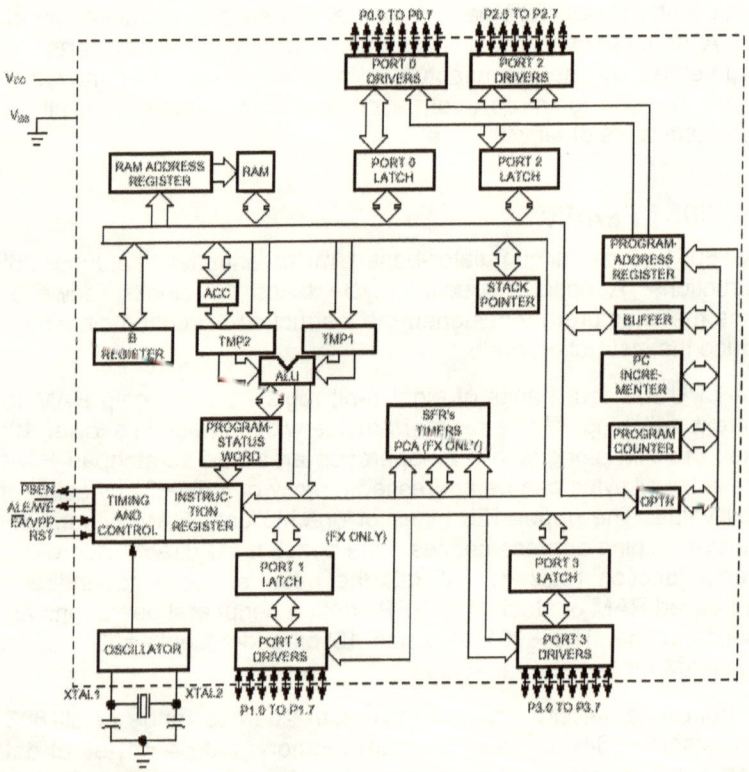

Figure 1   Block diagram 8051

To elucidate the differences in the derivatives, Figure 2 shows the block diagram of the C8051F0000 microcontroller from Cygnal [2].

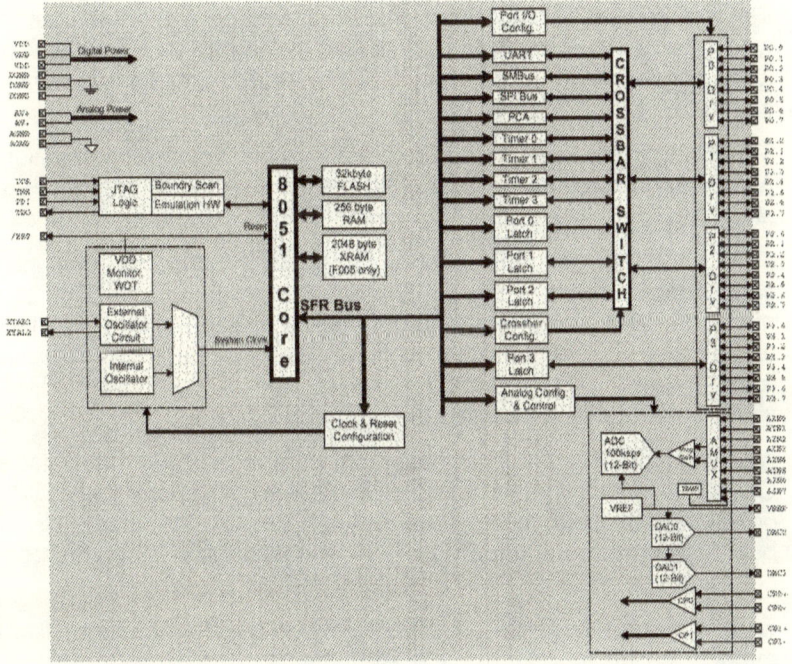

Figure 2   Block diagram C8051F0000

This is not the place to discuss the hardware aspects of the different derivatives of the 8051 family. The examples are meant to show that not all parts named 8051 are alike; the core is the same but the internal peripherals differ significantly.

Once you know the used hardware, you can organize the access to the resources of the chosen microcontroller.

1.2 AVR Family

Since Atmel's AVR microcontrollers were introduced to the market only a few years ago, they are not so well known as the 8051 controllers. Therefore, this interesting microcontroller family should be described in more detail.

Atmel's AVR microcontrollers use a new RISC architecture which has been developed to take advantage of the semiconductor integration and software capabilities of the 1990's. The resulting microcontrollers offer the highest MIPS/mW capability available in the 8-bit microcontrollers market today.

The architecture of the AVR microcontrollers was designed together with C-language experts to ensure that the hardware and software work hand-in-hand to develop a highly efficient, high-performance code.

To optimize the code size, performance and power consumption, AVR microcontrollers have big register files and fast one-cycle instructions.

The family of AVR microcontrollers includes differently equipped controllers - from a simple 8-pin microcontroller up to a high-end microcontroller with a large internal memory. The Harvard architecture addresses memories up to 8 MB directly. The register file is "dual mapped" and can be addressed as part of the on-chip SRAM, whereby fast context switches are possible.

All AVR microcontrollers are based on Atmel's low-power nonvolatile CMOS technology. The on-chip in-system programmable (ISP), downloadable flash memory permits devices on the user's circuit board to be reprogrammed via SPI or with the help of a conventional programming device.

By combining the efficient architecture with the downloadable flash memory on the same chip, the AVR microcontrollers represent an efficient approach to applications in the "Embedded Controller" market.

Table 1 shows an overview of the devices available today, including the configuration of the internal memory and I/O. Further information can be found on Atmel's web site [http://www.atmel.com] and in the literature [3].

Device	Flash [KB]	EEPROM	SRAM	I/O Pins
ATtiny11	1	0	0	6
ATtiny12	1	64	0	6
ATtiny22	2	128	90	5
AT90S1200	1	64	0	15
AT90S2313	2	128	128	15
AT90S2323	2	128	128	3
AT90S2343	2	128	128	5
AT90S2333	2	128	128	20
AT90S4414	4	256	256	32
AT90S4433	4	256	128	20
AT90S4434	4	256	256	32
AT90S8515	8	512	512	32
AT90S8534	8	512	256	15
AT90S8535	8	512	512	32
ATmega603	64	2K	4K	48
ATmega103	128	4K	4K	48

Table 1 AVR microcontrollers and their resources

The internal resources of the AVR microcontrollers will be considered with AT90S8515 used as an example. Figure 3 shows the block diagram of an AT90S8515.

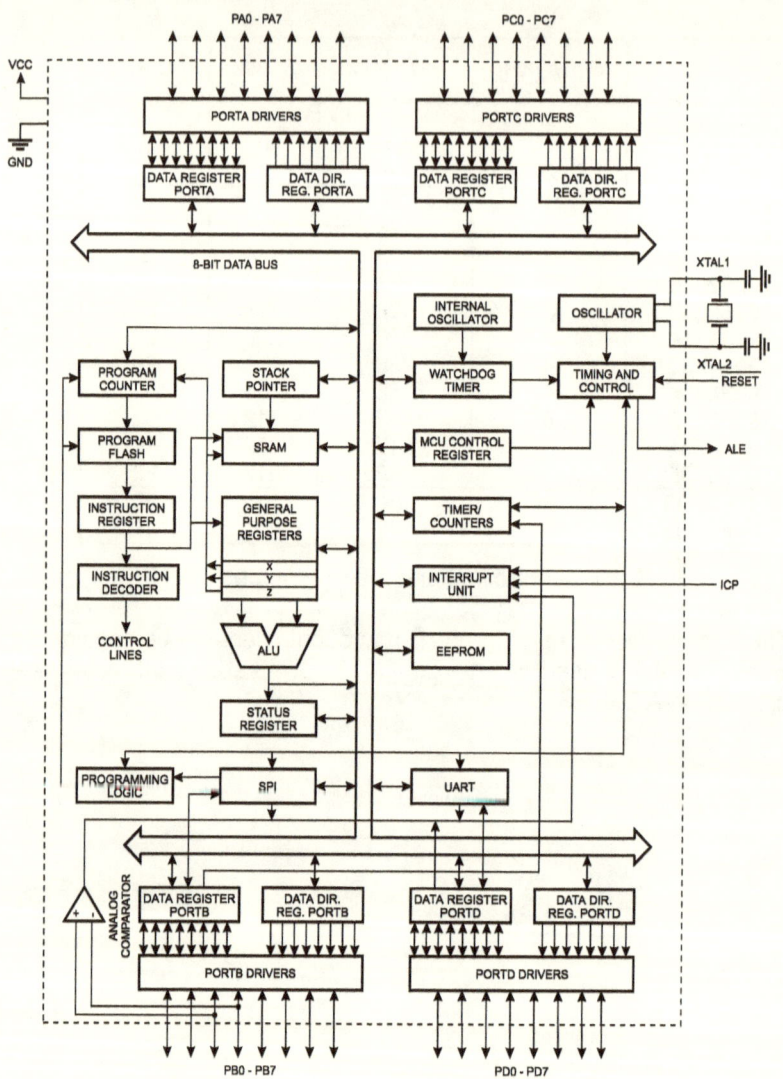

Figure 3 Block diagram AT90S8515

The I/O storage area covers 64 addresses for the peripheral device functions of the CPU, like control registers, Timer/Counter and other I/O functions. Figure 4 shows memory maps of the AT90S8515 program and data memory.

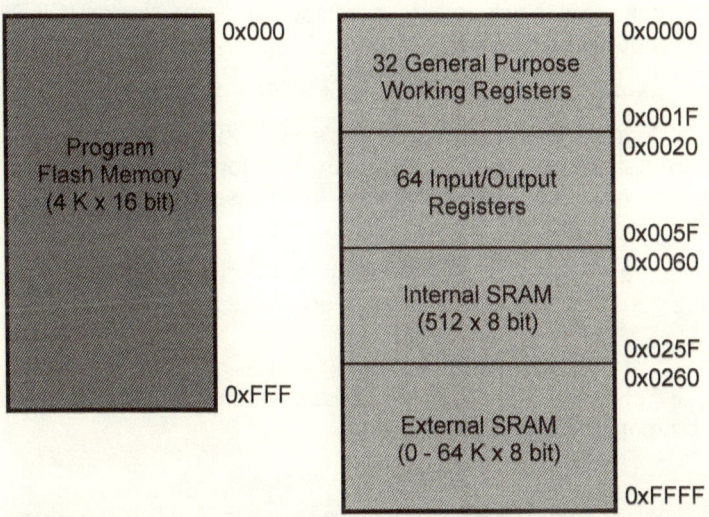

Figure 4
Memory maps for program and data memory for AT90S8515

The AVR microcontrollers make use of a Harvard structure with separate memories and busses for programs and data

A flexible interrupt module has its control register in the I/O memory area, too. All interrupts have separate interrupt vectors in an interrupt vector table at the beginning of the program memory. The priority level of each interrupt vector is dependent on its position in the interrupt vector table. The higher the priority of a respective interrupt, the lower is the address of the interrupt vector. All interrupts are maskable and can be enabled or disabled by a Global Interrupt Enable/Disable.

To get an impression of the available peripheral functions, the peripheral functions of the AT90S8515 will be listed here in brief as an example.

Timer/Counter

One 8-bit and one 16-bit Timer/Counter are available in conjunction with a flexible 10-bit prescaler for different timer and counter applications.

Both Timer/Counter units can operate independently as a timer with internal clock or as a counter with external triggering. The prescaler divides the internal clock into four selectable timer clocks (CK/8, CK/64, CK/256 and CK/1024).

The 8-bit Timer/Counter is a simple UpCounter.

The 16-bit Timer/Counter is more complex and supports two Output Compare functions and one Input Capture function. Furthermore, it is possible to use the Timer/Counter for Pulse-Width-Modulation (PWM).

The Watchdog Timer is clocked by a separate on-chip oscillator. The Watchdog period can be selected between 16 ms and 2048 ms.

SPI

The Serial Peripheral Interface (SPI) allows synchronous serial high-speed communication.

UART

A comfortable Universal Asynchronous Receiver/Transmitter (UART) allows flexible asynchronous serial communication.

Analog Comparator

The Analog Comparator compares voltages at two pins.

I/O Ports

The AT90S8515 has four I/O ports, which can be operate as digital input or output controlled by the Data Direction Register (DDR). As shown in Figure 5, most pins have alternative functions.

Comparing the pin configuration of the AVR microcontrollers and that of the 8051 microcontroller family reveals one objective of this new microcontroller family.

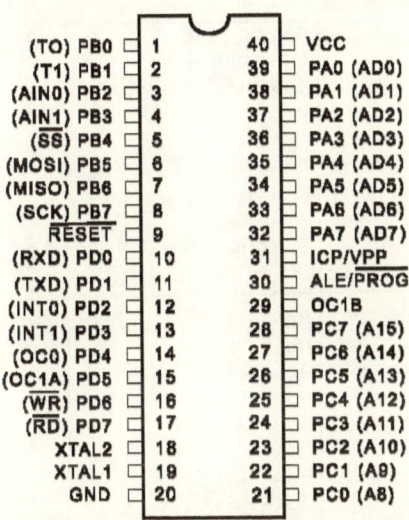

Figure 5   Pin configuration AT90S8515

All I/O ports are bidirectional with individually selectable Pull-up resistors. The outputs can drop to 20 mA so that LEDs can be directly driven.

The AVR microcontrollers support a high-voltage (12 V) parallel programming mode and a low-voltage serial programming mode. The serial programming mode via SPI provides a convenient way to download programs and data into the device inside the user's system.

To get an impression of the instruction set of the AVR microcontrollers, Table 2 explains all instructions in a compact form.

Mnemonics	Description		Cycles
ARITHMETIC AND LOGIC INSTRUCTIONS			
ADD Rd, Rr	Add without Carry	Rd ← Rd + Rr	1
ADC Rd, Rr	Add with Carry	Rd ← Rd + Rr + C	1
ADIW Rd, K	Add Immediate to Word	Rd+1:Rd ← Rd+1:Rd + K	2
SUB Rd, Rr	Subtract without Carry	Rd ← Rd - Rr	1
SUBI Rd, K	Subtract Immediate	Rd ← Rd - K	1
SBC Rd, Rr	Subtract with Carry	Rd ← Rd - Rr - C	1
SBCI Rd, K	Subtract Immediate with Carry	Rd ← Rd - K - C	1
SBIW Rd, K	Subtract Immediate from Word	Rd+1:Rd ← Rd+1:Rd - K	2
AND Rd, Rr	Logical AND	Rd ← Rd • Rr	1
ANDI Rd, K	Logical AND with Immediate	Rd ← Rd • K	1
OR Rd, Rr	Logical OR	Rd ← Rd v Rr	1
ORI Rd, K	Logical OR with Immediate	Rd ← Rd v K	1
EOR Rd, Rr	Exclusive OR	Rd ← Rd ⊕ Rr	1
COM Rd	One's Complement	Rd ← $FF - Rd	1
NEG Rd	Two's Complement	Rd ← $00 - Rd	1
SBR Rd,K	Set bit(s) in Register	Rd ← Rd v K	1
CBR Rd,K	Clear bit(s) in Register	Rd ← Rd • ($FFh - K)	1
INC Rd	Increment Rd ← Rd + 1	Rd ← Rd + 1	1
DEC Rd	Decrement	Rd ← Rd - 1	1
TST Rd	Test for Zero or Minus	Rd ← Rd • Rd	1
CLR Rd	Clear Register	Rd ← Rd ⊕ Rd	1
SER Rd	Set Register	Rd ← $FF	1
MUL Rd,Rr	Multiply Unsigned	R1, R0 ← Rd × Rr	2
BRANCH INSTRUCTIONS			
RJMP k	Relative Jump	PC ← PC + k + 1	2
IJMP	Indirect Jump to (Z)	PC ← Z	2
JMP k	Jump	PC ← k	3
RCALL k	Relative Call Subroutine	PC ← PC + k + 1	3
ICALL	Indirect Call to (Z)	PC ← Z	3
CALL k	Call Subroutine	PC ← k	4
RET	Subroutine Return	PC ← STACK	4
RETI	Interrupt Return	PC ← STACK	4
CPSE Rd,Rr	Compare, Skip if Equal if (Rd = Rr)	PC ← PC + 2 or 3	1 / 2
CP Rd,Rr	Compare	Rd - Rr	1
CPC Rd,Rr	Compare with Carry	Rd - Rr - C	1
CPI Rd,K	Compare with Immediate	Rd - K	1
SBRC Rr, b	Skip if bit in Register Cleared	if (Rr(b)=0) PC ← PC + 2 or 3	1 / 2
SBRS Rr, b	Skip if bit in Register Set	if (Rr(b)=1) PC ← PC + 2 or 3	1 / 2
SBIC P, b	Skip if bit in I/O Register Cleared	if(I/O(P,b)=0) PC ← PC + 2 or 3	2 / 3
SBIS P, b	Skip if bit in I/O Register Set	If(I/O(P,b)=1) PC← PC + 2 or 3	2 / 3
BRBS s, k	Branch if Status Flag Set	if (SREG(s) = 1) then PC←PC+k + 1	1 / 2
BRBC s, k	Branch if Status Flag Cleared	if (SREG(s) = 0) then PC←PC+k + 1	1 / 2

BREQ k	Branch if Equal	if (Z = 1) then PC ← PC + k + 1	1 / 2
BRNE k	Branch if Not Equal	if (Z = 0) then PC ← PC + k + 1	1 / 2
BRCS k	Branch if Carry Set	if (C = 1) then PC ← PC + k + 1	1 / 2
BRCC k	Branch if Carry Cleared	if (C = 0) then PC ← PC + k + 1	1 / 2
BRSH k	Branch if Same or Higher	if (C = 0) then PC ← PC + k + 1	1 / 2
BRLO k	Branch if Lower	if (C = 1) then PC ← PC + k + 1	1 / 2
BRMI k	Branch if Minus	if (N = 1) then PC ← PC + k + 1	1 / 2
BRPL k	Branch if Plus	if (N = 0) then PC ← PC + k + 1	1 / 2
BRGE k	Branch if Greater or Equal, Signed	if (N $\oplus$ V= 0) then PC ← PC+ k + 1	1 / 2
BRLT k	Branch if Less Than, Signed	if (N $\oplus$ V= 1) then PC ← PC + k + 1	1 / 2
BRHS k	Branch if Half Carry Flag Set	if (H = 1) then PC ← PC + k + 1	1 / 2
BRHC k	Branch if Half Carry Flag Cleared	if (H = 0) then PC ← PC + k + 1	1 / 2
BRTS k	Branch if T Flag Set	if (T = 1) then PC ← PC + k + 1	1 / 2
BRTC k	Branch if T Flag Cleared	if (T = 0) then PC ← PC + k + 1	1 / 2
BRVS k	Branch if Overflow Flag is Set	if (V = 1) then PC ← PC + k + 1	1 / 2
BRVC k	Branch if Overflow Flag is Cleared	if (V = 0) then PC ← PC + k + 1	1 / 2
BRIE k	Branch if Interrupt Enabled	if (I = 1) then PC ← PC + k + 1	1 / 2
BRID k	Branch if Interrupt Disabled	if (I = 0) then PC ← PC + k + 1	1 / 2
DATA TRANSFER INSTRUCTIONS			
MOV Rd, Rr	Copy Register	Rd ← Rr	1
LDI Rd, K	Load Immediate	Rd ← K	1
LDS Rd, k	Load Direct from SRAM	Rd ← (k)	3
LD Rd, X	Load Indirect	Rd ← (X)	2
LD Rd, X+	Load Indirect and Post-Increment	Rd ← (X), X ← X + 1	2
LD Rd, -X	Load Indirect and Pre-Decrement	X ← X - 1, Rd ← (X)	2
LD Rd, Y	Load Indirect	Rd ← (Y)	2
LD Rd, Y+	Load Indirect and Post-Increment	Rd ← (Y), Y ← Y + 1	2
LD Rd, -Y	Load Indirect and Pre-Decrement	Y ← Y - 1, Rd ← (Y)	2
LDD Rd,Y+q	Load Indirect with Displacement	Rd ← (Y + q)	2
LD Rd, Z	Load Indirect	Rd ← (Z)	2

Mnemonic	Description	Operation	Clocks
LD Rd, Z+	Load Indirect and Post-Increment	Rd ← (Z), Z ← Z+1	2
LD Rd, -Z	Load Indirect and Pre-Decrement	Z ← Z - 1, Rd ← (Z)	2
LDD Rd, Z+q	Load Indirect with Displacement	Rd ← (Z + q)	2
STS k, Rr	Store Direct to SRAM	Rd ← (k)	3
ST X, Rr	Store Indirect	(X) ← Rr	2
ST X+, Rr	Store Indirect and Post-Increment	(X) ← Rr, X ← X + 1	2
ST -X, Rr	Store Indirect and Pre-Decrement	X ← X - 1, (X) ← Rr	2
ST Y, Rr	Store Indirect	(Y) ← Rr	2
ST Y+, Rr	Store Indirect and Post-Increment	(Y) ← Rr, Y ← Y + 1	2
ST -Y, Rr	Store Indirect and Pre-Decrement	Y ← Y - 1, (Y) ← Rr	2
STD Y+q,Rr	Store Indirect with Displacement	(Y + q) ← Rr	2
ST Z, Rr	Store Indirect	(Z) ← Rr	2
ST Z+, Rr	Store Indirect and Post-Increment	(Z) ← Rr, Z ← Z + 1	2
ST -Z, Rr	Store Indirect and Pre-Decrement	Z ← Z - 1, (Z) ← Rr	2
STD Z+q,Rr	Store Indirect with Displacement	(Z + q) ← Rr	2
LPM	Load Program Memory	R0 ← (Z)	3
IN Rd, P	In Port	Rd ← P	1
OUT P, Rr	Out Port	P ← Rr	1
PUSH Rr	Push Register on Stack	STACK ← Rr	2
POP Rd	Pop Register from Stack	Rd ← STACK	2

BIT AND BIT-TEST INSTRUCTIONS

Mnemonic	Description	Operation	Clocks
LSL Rd	Logical Shift Left	Rd(n+1)←Rd(n), Rd(0)←0,C←Rd(7)	1
LSR Rd	Logical Shift Right	Rd(n)←Rd(n+1), Rd(7)←0,C←Rd(0)	1
ROL Rd	Rotate Left Through Carry	Rd(0)←C, Rd(n+1)←Rd(n),C←Rd(7)	1
ROR Rd	Rotate Right Through Carry	Rd(7)←C, Rd(n)←Rd(n+1),C←Rd(0)	1
ASR Rd	Arithmetic Shift Right	Rd(n) ← Rd(n+1), n=0..6	1
SWAP Rd	Swap Nibbles	Rd(3..0) ↔ Rd(7..4)	1
BSET s	Flag Set	SREG(s) ← 1	1
BCLR s	Flag Clear	SREG(s) ← 0	1
SBI P, b	Set bit in I/O Register	I/O(P, b) ← 1	2
CBI P, b	Clear bit in I/O Register	I/O(P, b) ← 0	2
BST Rr, b	bit Store from Register to T	T ← Rr(b)	1
BLD Rd, b	bit load from T to Register	Rd(b) ← T	1
SEC	Set Carry	C ← 1	1
CLC	Clear Carry	C ← 0	1
SEN	Set Negative Flag	N ← 1	1
CLN	Clear Negative Flag	N ← 0	1

SEZ	Set Zero Flag	Z ← 1	1
CLZ	Clear Zero Flag	Z ← 0	1
SEI	Global Interrupt Enable	I ← 1	1
CLI	Global Interrupt Disable	I ← 0	1
SES	Set Signed Test Flag	S ← 1	1
CLS	Clear Signed Test Flag	S ← 0	1
SEV	Set Two's Complement Overflow	V ← 1	1
CLV	Clear Two's Complement Overflow	V ← 0	1
SET	Set T in SREG	T ← 1	1
CLT	Clear T in SREG	T ← 0	1
SEH	Set Half Carry Flag in SREG	H ← 1	1
CLH	Clear Half Carry Flag in SREG	H ← 0	1
NOP	No Operation	None	1
SLEEP	Sleep		1
WDR	Watchdog Reset		1

Table 2 Instruction Set of AVR microcontrollers

These introducing remarks on the AVR microcontrollers cannot of course replace a detailed study of the technical documentation of the manufacturer. Descriptions of the individual microcontrollers as well as application notes and program examples can be found on Atmel's web site [http://www.atmel.com]. The manufacturer's documentation is complemented by further publications [3][4].

2 BASCOM

BASCOM-AVR is not only a BASIC Compiler, but also a comfortable Integrated Development Environment (IDE) running under Windows 95 and Windows NT.

Such a development environment supports the whole process from coding and testing a program to programming the used microcontroller.

In this book the term BASCOM is used when no distinction must be made between BASCOM-8051 and BASCOM-AVR. In all cases where a distinction is necessary, a few changes only are required to make the program work with the other family of microcontrollers. This is one important advantage of high-level languages.

So as to prevent that work with BASCOM and the program examples in this book are mere dry homework, a demo of BASCOM-8051 or BASCOM-AVR can be used for first tests. These BASCOM demos can be downloaded free of charge from different URLs.

For proper installation of the required BASCOM IDE, make sure a printer is installed - the printer need not necessarily be used or connected.

The licence agreement must be accepted before one of the BASCOM IDEs is installed

2.1 BASCOM Demos

Over a link to the download area of the BASCOM developer MCS Electronics [http://www.mcselec.com] some files are available for download.

For download the BASCOM-8051 demo use this URL

http://www.mcselec.com/download_8051.htm

and for downloading the BASCOM-AVR demo use

http://www.mcselec.com/download_avr.htm .

On these download sites you will find the manuals as PDF and all information required for an upgrade to the commercial versions.

After extracting all downloaded files to a separate directory, there is a setup program for installation.

Installation starts as usual under Windows when this setup program is called.

After completion of the installation, the following files need to be installed on the PC. Figure 6 shows the files installed for BASCOM-AVR as an example. Inspecting the directory with the Explorer will show some more files there. These files will be explained later.

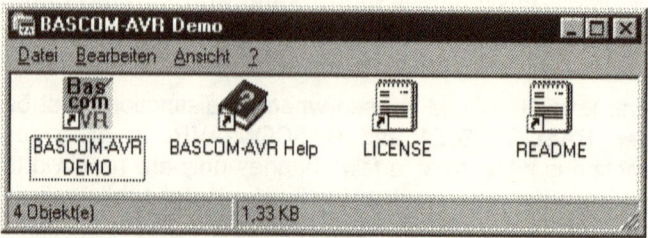

Figure 6   BASCOM-AVR Demo Files

As is common for most demo programs, some restrictions must be expected. The only restriction of both BASCOM demos is a reduced code size of 2 KB.

If the code size exceeds this limit after compilation, the compiler will generate error messages as shown in Figure 7.

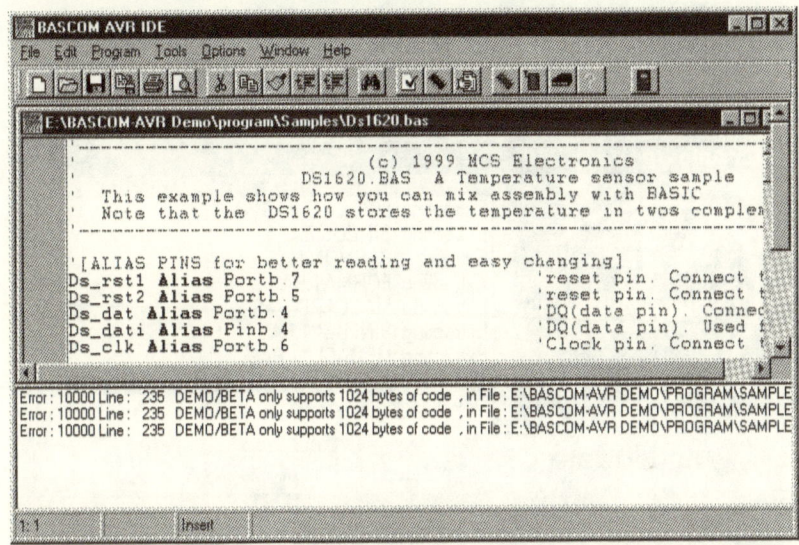

Figure 7   Error messages due to exceeding the restricted code size

2.2 BASCOM Commercial Versions

If you decide to buy the commercial version of the used BASCOM IDE, you may order it from http://www.mcselec.com or one of the local distributors. Downloading the files and ordering the license is done in next to no time. The license will be sent immediately by e-mail.

The installation of the commercial version does not differ from the procedure for the BASCOM demo. Start SetUp and follow the instructions of the SetUp program.

2.3 Update of BASCOM Commercial Versions

When a commercial version of BASCOM is installed, it can be updated when a new version is ready for downloading from MCS Electronic's web site. In the download area you will find a link to an AutoUpdate program.

Install this AutoUpdate program in your BASCOM-8051 or BASCOM-AVR subdirectory as you installed BASCOM-8051 or BASCOM-AVR before.

Figure 8 shows the downloading and extracting of updated files in an existing installation of BASCOM-AVR.

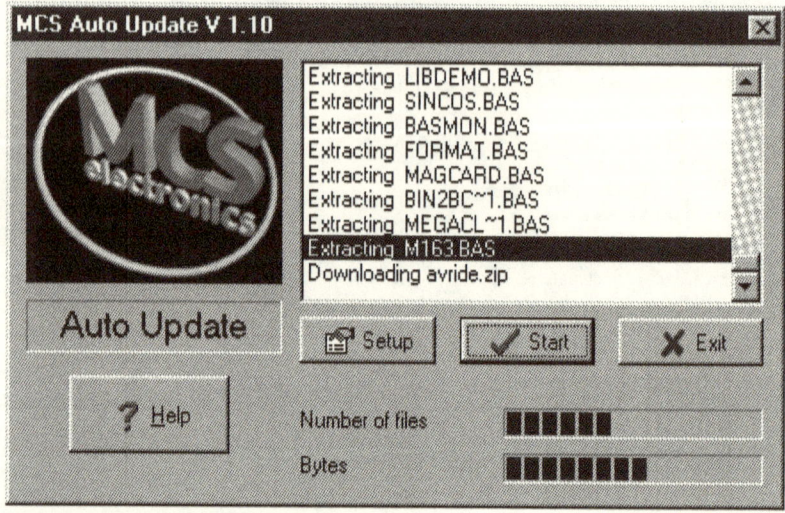

Figure 8   Update of BASCOM-AVR

If your installation is up-to-date then there is no need for an update. The AutoUpdate program detects this state automatically (Figure 9).

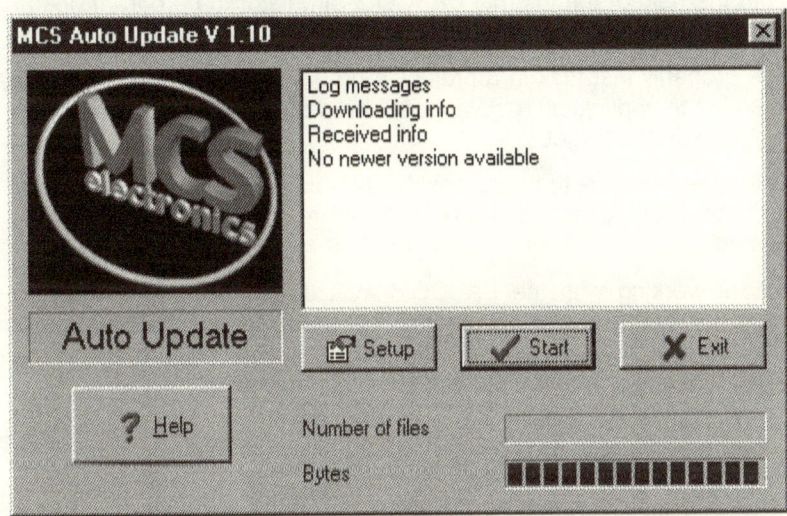

Figure 9 No newer version available

If you use the AutoUpdate program from time to time you will always have an actual installation of the used BASCOM IDE.

2.4 BASCOM Projects

2.4.1 Working on Projects

After the start of BASCOM you can create a new file by selecting *File>New* or open an existing file by selecting *File>Open*.

In the next step, check such BASCOM Options like device selection, baud rate, clock frequency and other relating options. A detailed explanation of these options will be given in the next chapter.

Now you may edit the BASIC source and compile it afterwards. As a rule, the compiler detects here the first errors and the program must be debugged.

The BASIC source must be edited as long as the compilation is without any errors. Normally, the process of editing, compiling and debugging needs to be repeated several times. It makes no sense to debug all errors in one step. Editing several typing errors in one step is no problem. But for more difficult errors, a separate compiler run checks the validity of the changes carried out. It is always easier to debug a localized error.

With the help of the internal BASCOM Simulator the program operation can be checked without any hardware.

The probably last task in a project is programming the device that is used in the application hardware, followed by an excessive test of the program on the target.

The project proves to be successful if these tests document a proper function in the target hardware. Otherwise, some steps must be repeated.

Before working with the BASCOM-AVR, the development environment will be described by means of a small program example; the next chapter describes the BASCOM options important to the BASCOM environment used and the target hardware.

2.4.2 BASCOM Options

Each BASCOM offers a lot of options that must be defined by selection in the Option menu. The options should be selected at the beginning of a project and saved. Later changes of this setup will then only be required for details.

The following description applies to BASCOM-AVR. In BASCOM-8051, selecting the various options is quite similar.

In the first step, the used microcontroller is defined by selecting *Options>Compiler>Chip*. Let us use here an AT90S8515 without external RAM. Figure 10 shows the parameters. On the right side you can see the available memory of the selected microcontroller.

Each parameter in a function needs two bytes of stack. the stack size shows the number of reserved bytes for the stack. The value 32 is default and remains unchanged here.

Local variables are saved in a frame. The default value is 50 and remains unchanged, too.

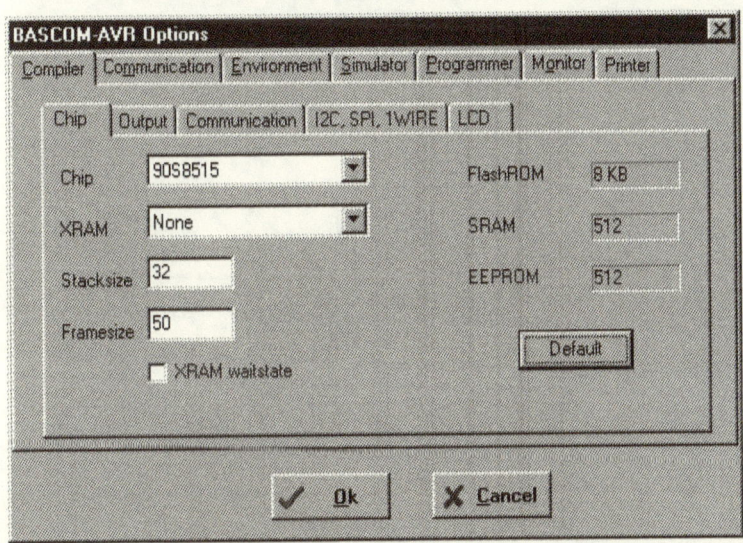

Figure 10   Selection of a device and external memory

The compiler generates many files selectable by *Options> Compiler>Output*. Figure 11 shows the possibilities for selection.

In dependence of the used programmer, Bin files and/or Hex files will be generated. The compiler itself needs the debug file. The report file reports all parameters and memory allocations. The error file documents all errors occurring during compilation.

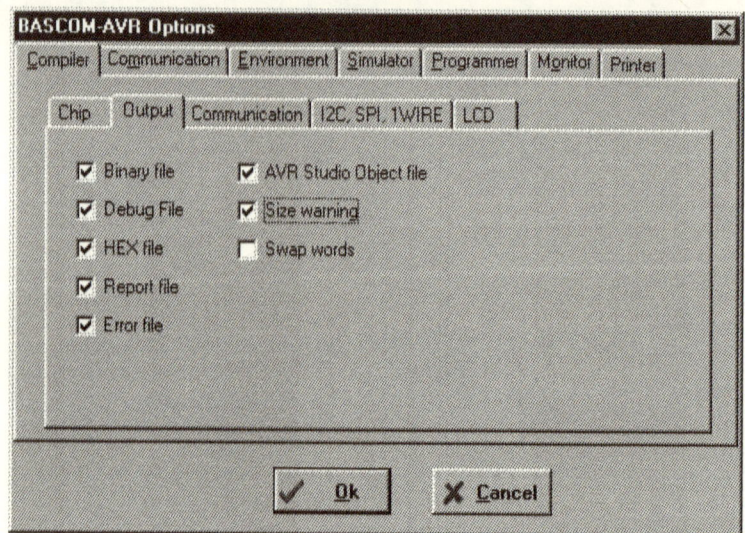

Figure 11   Selection of files to be generated

To simplify matters, all files on the left side should be selected.

For simulations with AVR Studio (AVR only), the related object file is required. Activating Size warning reports an exceeding of the available program memory. The last option can be very helpful.

Some programmers require Bin or Hex files with swapped LSB and MSB. In this case, activate the Swap Words option.

The baud rate of serial communication (RS232) depends on the clock frequency of the microcontroller. The clock frequency and desired baud rate can be selected from menu **Options> Compiler>Communication**. Figure 12 shows the parameter input. The error field shows the deviation of the generated baud rate.

It is very important to keep this deviation within defined limits as otherwise communication errors may occur.

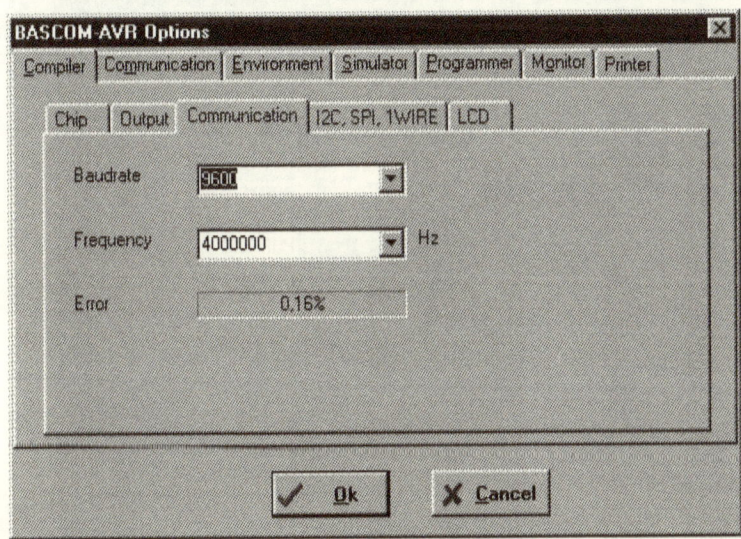

Figure 12 Selection of baud rate and oscillator frequency

In addition to serial communication according to RS232, BASCOM supports I^2C, SPI and 1-Wire data transfer. As Figure 13 shows, the menu *Options>Compiler>I2C, SPI, 1WIRE* allows the allocation of pins to the respective lines. At this time at the latest, a wiring diagram or schematic of the target hardware is required.

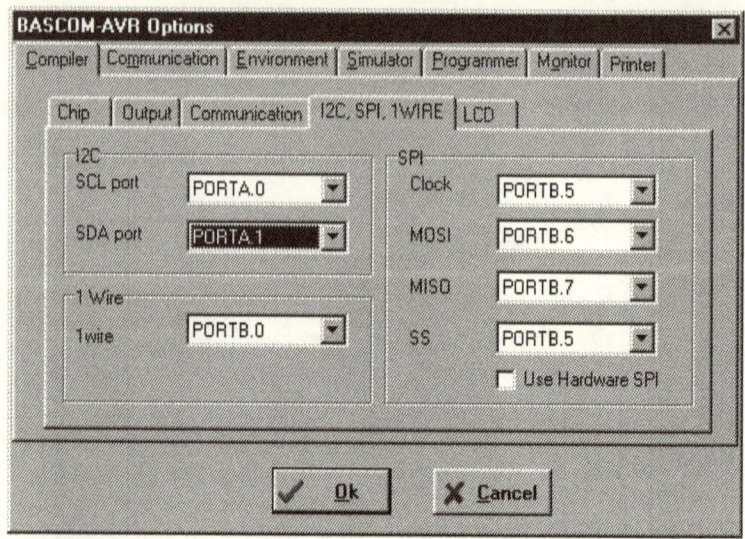

Figure 13 Selection of pins for serial communication

From menu **Options>Compiler>LCD** an LCD can be connected to the selected pins. Figure 14 shows the input of the required parameters.

For BASCOM-AVR there are different methods for controlling an LCD. If the microcontroller has an external RAM, then the LCD can be connected to the data bus. The address bus controls lines E and RS. The following connections are required in the bus mode.

AT90Sxxxx	A15	A14	D7	D6	D5	D4	D3	D2	D1	D0
8-bit Mode	E	RS	db7	db6	db5	db4	db3	db2	db1	db0
4-bit Mode	E	RS	db7	db6	db5	db4	-	-	-	-

For BASCOM-8051 and BASCOM-AVR it is possible to assign any pin of the microcontroller to the LCD pins. Usually, the 4-bit mode will be used (four data lines).

When defining user-specific characters, bit-maps are assigned to printable characters. This process is very simple and is supported by the LCD designer. Using the option "Make upper 3 bit 1 in LCD Designer" the bit-maps can be influenced as shown.

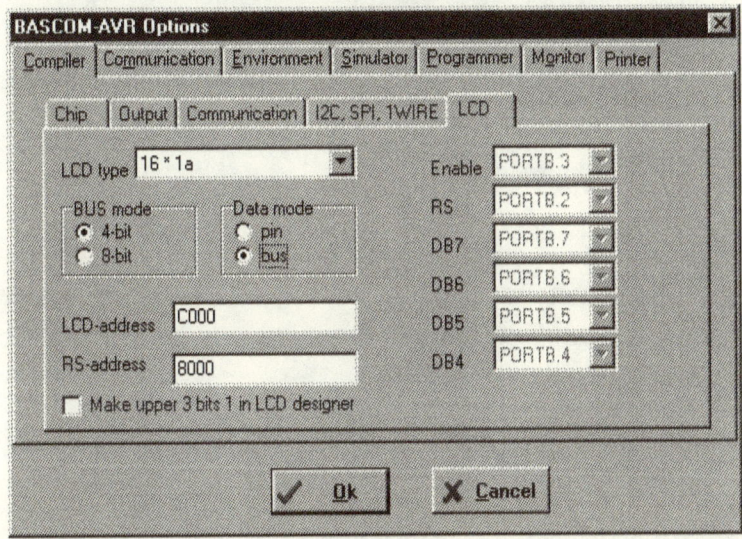

Figure 14 BASCOM-AVR LCD SetUp

When communicating from the PC with the target hardware, the parameters of the terminal emulator must be coordinated with the interface parameters of the target hardware. As is shown in Figure 15, these parameters can be input via the menu **Options> Communication**.

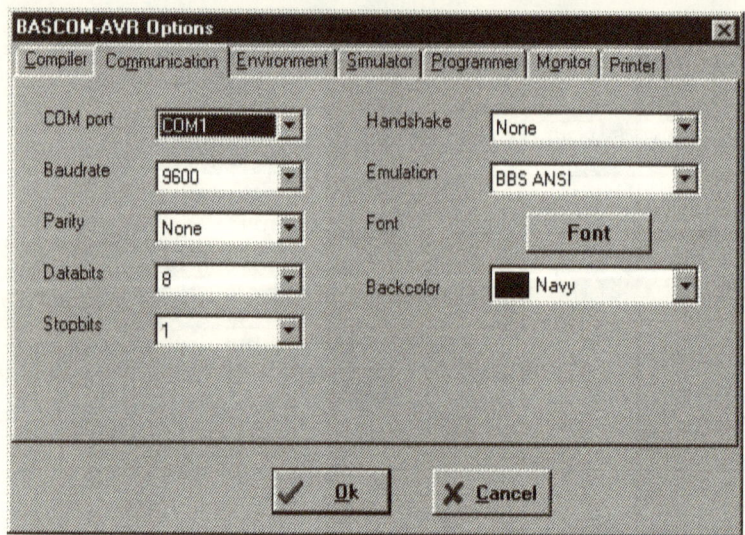

Figure 15   Parameter selection for terminal emulator

The editor features can be adapted as preferred. Figure 16 shows the setup options; they are selectable via menu *Options> Environment* .

As experience shows, the setup can be used as default for the first time. Any changes can be made later when you are more familiar with the editing of source text.

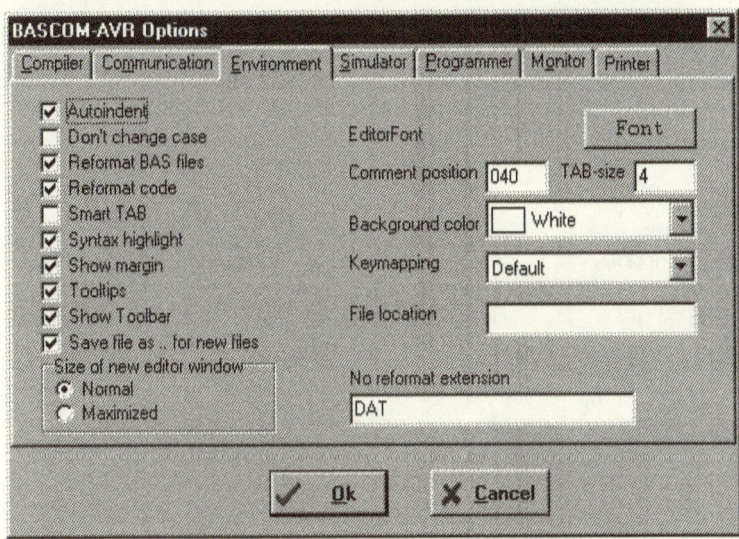

Figure 16 Selection of editor options

In BASCOM-AVR you can choose the internal simulator or AVR Studio for simulation. In menu *Options>Simulator* the AVR Studio can be linked to BASCOM-AVR. Figure 17 shows the link to AVRStudio.exe in path D:\Programme\AVRSTUD\, which is specific to the author's system.

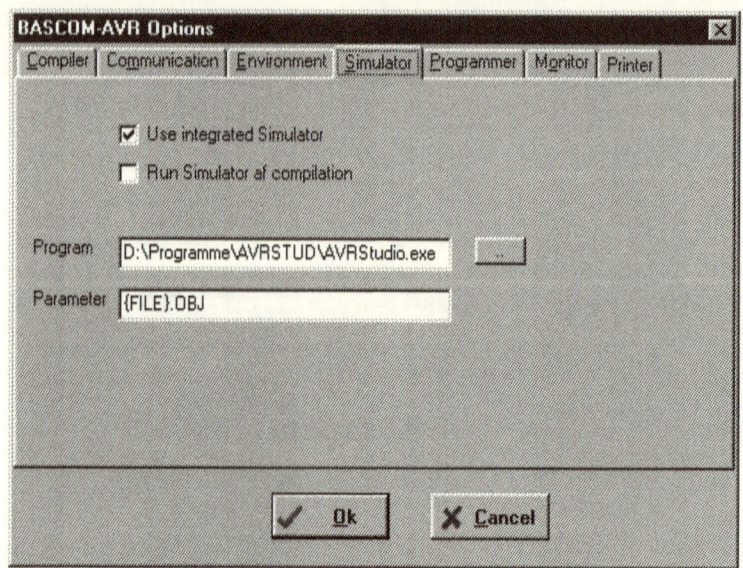

Figure 17 Selection of a simulator

The last important step is the selection of a programmer via menu **Options>Programmer**. Figure 18 shows this selection.

In this case, the AVR ISP Programmer was selected because most BASCOM-AVR program examples described here used the MCU00100 evaluation board as a hardware platform. Basically, the use of an external programmer is possible.

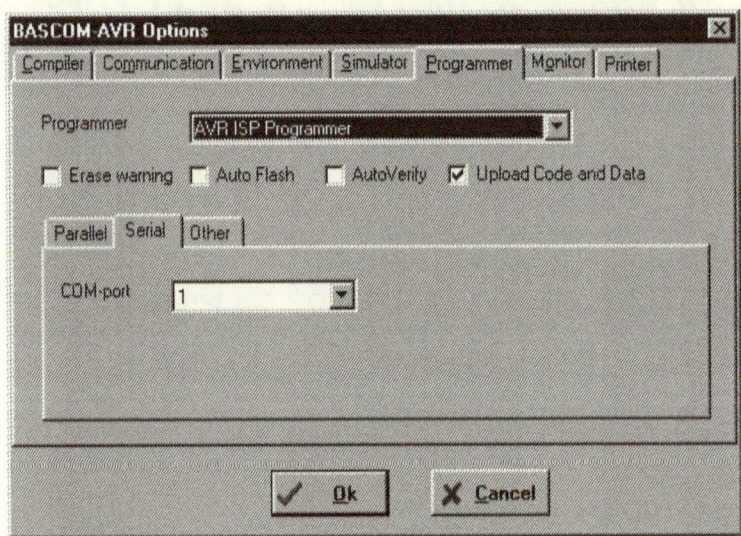

Figure 18   Selection of a programmer

2.5 BASCOM Tools

BASCOM IDE includes some important tools. The simulator and programmer have already been mentioned.

Further tools are

- a Terminal Emulator for communication with the serial interface of the target hardware,
- an LCD Designer supporting the design of customer-specific characters for a connected character LCD
- a library manager supporting the management of libraries and
- for BASCOM-805,1, a Graphic BMP Converter intended to convert BMP files into BASCOM Graphic Files (BGF) for display by a Graphic LCD.

2.5.1 Simulation

BASCOM-8051 and BASCOM-AVR have their own internal simulator. A simple program example describes the use of the simulator in both BASCOM IDEs.

The program to be simulated controls an alphanumeric LCD of two lines of 16 characters each. Listing 1 shows the source text.

```
$sim          ' for simulation only otherwise comment

Dim A As Byte

M1:
   A = Waitkey()

   If A = 27 Then Goto M2
   Cls
   Upperline
   Lcd A
   Lowerline
   Lcd Hex(a)
   Print Chr(a)
   Goto M1
M2:
   End
```

Listing 1   LCD Test (LCD.BAS)

Clicking **Program>Simulate** or **F2** starts the Simulator and the simulation window opens up.

Figure 19 shows the simulation window of BASCOM-8051 and Figure 20 that of BASCOM-AVR.

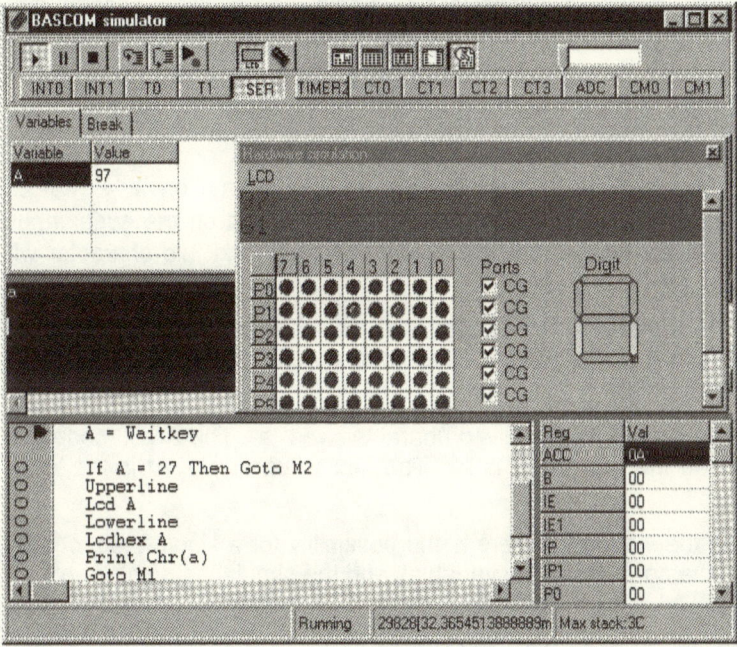

Figure 19 BASCOM-8051 Simulator

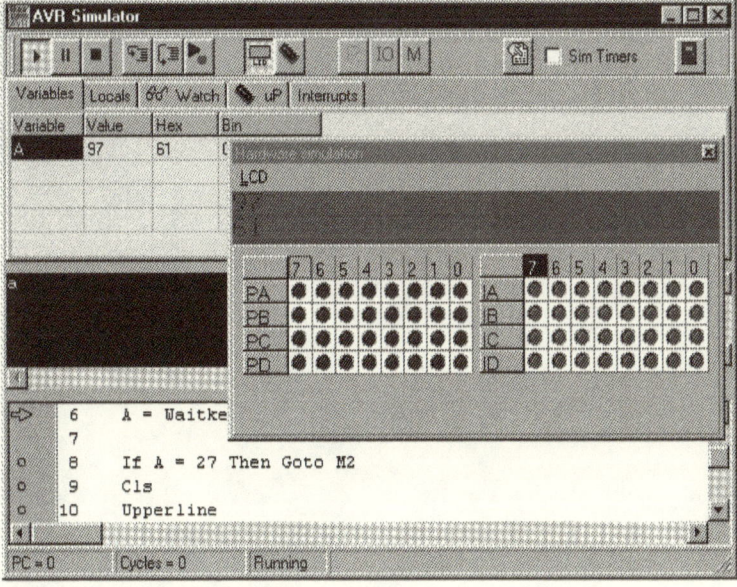

Figure 20 BASCOM-AVR Simulator

The program instructions can be seen at the bottom of the window. A terminal window is placed in the middle, and a watch window presenting the contents of the variables on top.

In the example, the control of an LCD is simulated. For the purpose, the LCD windows were opened. As can be seen, the LCD windows differ.

After program start the program runs until instruction `a = Waitkey()`, and waits for a character to be received on the serial input. Key in a character from the terminal window and this character will be read by the program.

If the received character is not ESC, its ASCII code will be displayed in the upper row of the LCD and its hex value in the lower row of the simulated LCD.

In the example, the received character was "a". The ASCII code displayed in the upper line is 92. The hex value displayed in the lower line is 61.

During the simulation there is the possibility for a single-step changing of the contents of the variables and the simulation of interrupts.

2.5.2 Terminal Emulator

The Terminal Emulator is used for communication with the serial interfaces of the target hardware.

Listing 2 shows a simple test program. The program waits until it receives one character, echoes this character, and adds some characters for commentary purposes.

```
Dim A As Byte

Do    A = Inkey()                'get   value from serial port

      If A > 0 Then              'we got something
      Print "Received from serial port:"
      Print "ASCII Code " ; A ;
      Print " = Character " ; Chr(a)
      End If

Loop Until A = 27

End
```

Listing 2   Test of serial communication (SERIAL.BAS)

To start the Terminal Emulator, click *Tools>Terminal emulator* or press *Ctrl+T*. Figure 21 shows the open terminal window. The parameters for communication can be selected via menu *Options>Communications*; they are shown in the status line.

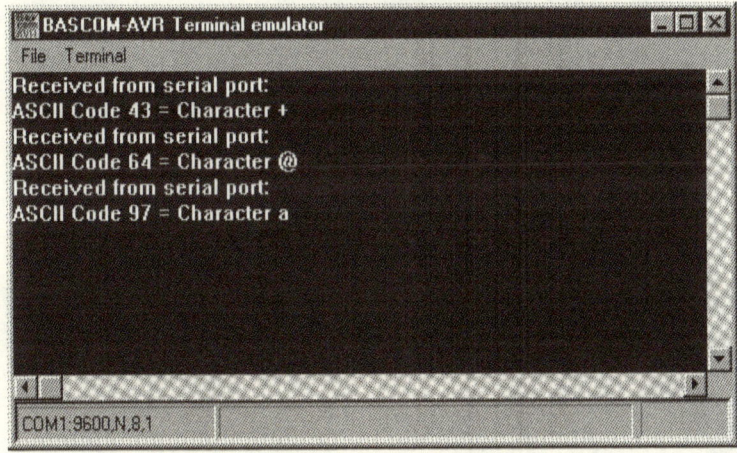

Figure 21   BASCOM-AVR Terminal Emulator

If a character is sent to the target hardware by typing this character in the PC's keyboard, then the program checks the received character (A = inkey()) and sends back a comment and the results of some operations (Print ...) until the ESC key is pressed and the program stops.

The Terminal Emulator can be used for testing all communication tasks of the serial interface of the used microcontroller.

2.5.3 LCD Designer

The LCD Designer is useful for defining customer-specific characters displayed on an alphanumeric LCD. All alphanumeric LCDs, working with Hitachi's LCD controller HD 44780 or a compatible, allow custom-specific characters to be defined.

Figure 22 shows three characters which are defined as custom-specific characters and tested.

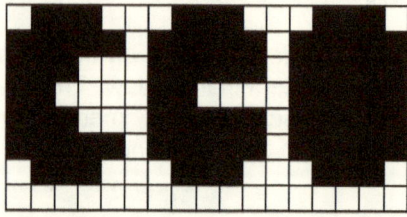

Figure 22

The first character is used to demonstrate custom-specific character definition with the help of the LCD Designer.

By *Tools>LCD Designer* or *Ctrl+L,* the LCD designer is started (Figure 23).

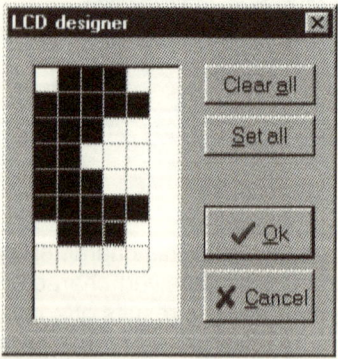

Figure 23
Custom-specific character definition in LCD Designer

The pixels in the 8x5 matrix can be set or cleared. The lowest pixel line, though reserved for the display of the LCD cursor, can be used.

By pressing OK the character is defined and the respective instruction is written in the source text window.

For the time being, the designation of the character is provided with a question mark which must be replaced by a character (or a variable) within the range from 0 to 7.

Figure 24 shows the entry in the source text completed by a constant of 1 as the name for this first user-specific character.

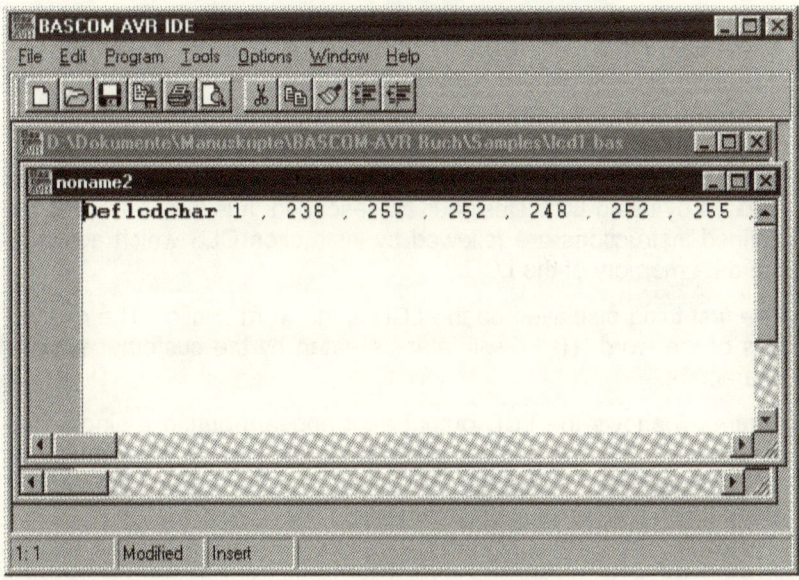

Figure 24 Instruction generated by the LCD designer

A small program (Listing 3) supports the test of these user-specific character indications.

```
Deflcdchar 1 , 238 , 255 , 252 , 248 , 252 , 255 , 238 , 224
Deflcdchar 2 , 238 , 255 , 255 , 248 , 255 , 255 , 238 , 224
Deflcdchar 3 , 238 , 255 , 255 , 255 , 255 , 255 , 238 , 224

Cls

Config Lcd = 16 * 1
Lcd "Hello "
Home
Lcd Chr(1)
Home
Lcd " " ; Chr(2)
Home
Lcd " " ; Chr(3)
Home
Lcd "    " ; Chr(1)
Home
Lcd "     " ; Chr(2)
Home
Lcd "      " ; Chr(3)
```

Listing 3   Customer-specific characters (LCD1.BAS)

At the beginning of the program there are three character definitions created by using LCD Designer as described. It is important that the defined instructions are followed by instruction CLS which activates the data memory of the LCD.

The first thing displayed on the LCD is the word "Hello". The characters of the word "Hello" will later be eaten by the customer-specific characters.

Figure 25 shows the LCD output as it appears during a single-step simulation one after another. Several hardcopies of the Simulator's LCD window were cascaded one below the other so that the various steps taken can be seen very clearly.

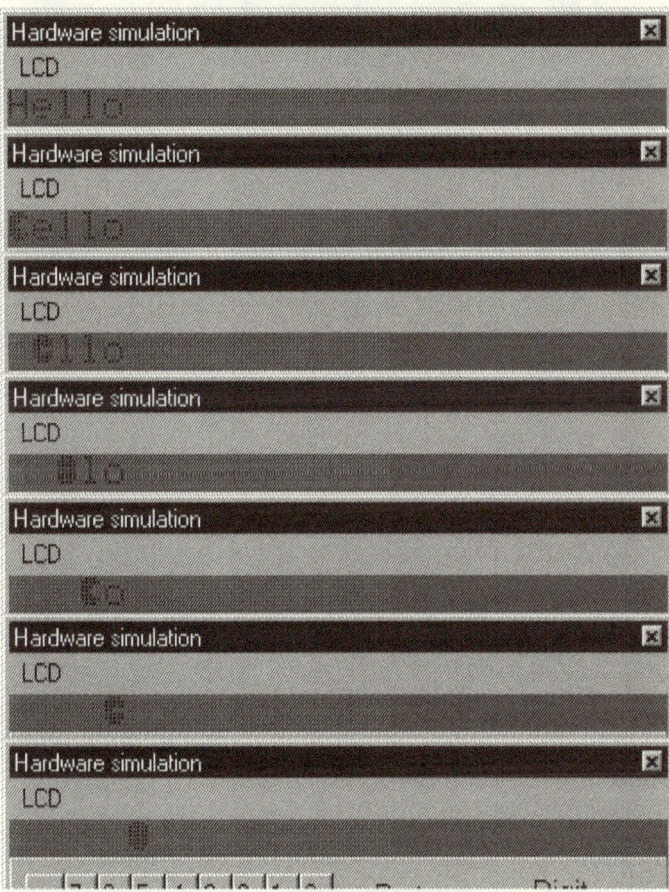

Figure 25 LCD output in Simulator

2.5.4 Library Manager

A library contains assembler routines which can be accessed from a program. The Library Manager supports the administration and modification of such a library.

Figure 26 shows routines of the library MCS.LIB for BASCOM-8051.

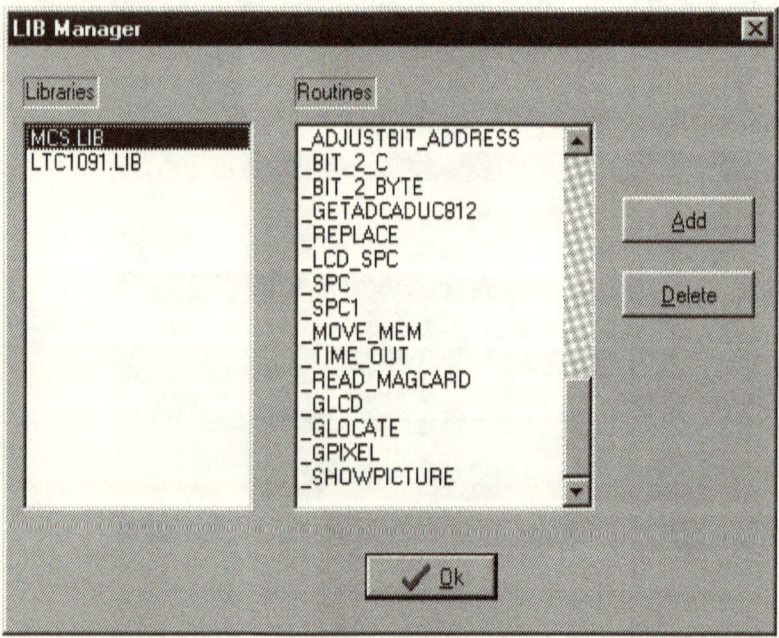

Figure 26   BASCOM-8051 LIB Manager

Figure 27 shows routines of the library MCS.LIB for BASCOM-AVR.

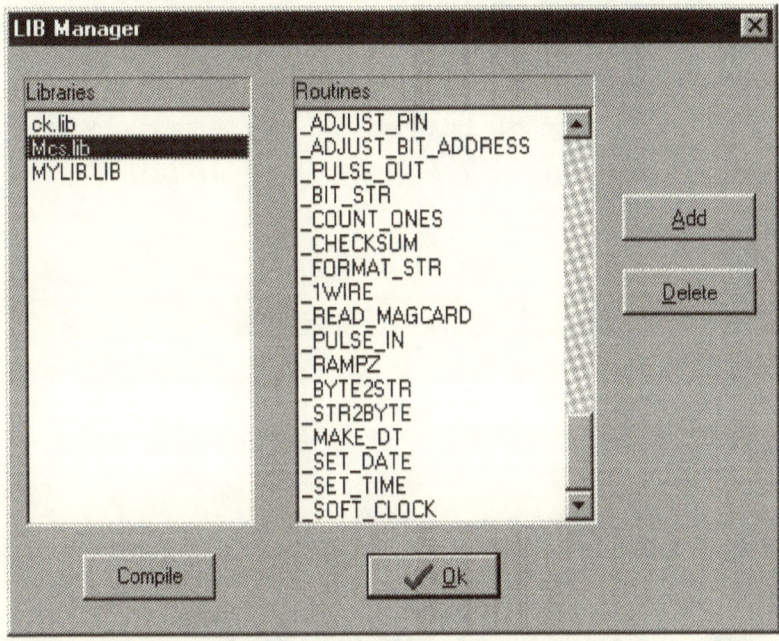

Figure 27 BASCOM-8051 LIB Manager

The libraries will be searched when a used routine is declared with the directive $EXTERNAL. The library search order is the same as the order of the names of the libraries. Library MCS.LIB included in both BASCOM IDEs is always the last library searched. There is no need to specify MCS.LIB by the directive $LIB.

Since MCS.LIB is always the last library searched, routines with the same name but a different function can be included in private libraries. Because of the search order, that routine is found first and thus redefines the definition in MCS.LIB.

To change the predefined routines in MCS.LIB, copy and rename MCS.LIB and edit the routines to be changed. It is als possible to create private libraries. Listing 4 shows a BlockMove routine for BASCOM-AVR in library CK.LIB.

```
Copyright = Claus Kuehnel
WWW = http://www.ckuehnel.ch
Email = avr@ckuehnel.ch
Comment = Avr Compiler Library
Libversion = 1.00
Date = 19.01.2000
Statement = No Source Code From The Library May Be Distributed In
Any Form
Statement = Of Course This Does Not Applie For The Compiled Code
When You Have A Bascom -  Avr License
History = No Known Bugs.
History =

...

[_blockmove]
_blockmove:
 ld   _temp1,Z+            ;get data from BLOCK1
 st   X+,_temp1            ;store data to BLOCK2
 dec  _temp2               ;
 brne _blockmove           ;if not done, loop more
 ret                       ;return
[end]
```

Listing 4   Library CK.LIB

A library is a simple text file. Each editor can be used for making changes. By means of the BASCOM internal editor, a library can be edited in the same way as a BASIC source file.

The header contains some useful information.

Each routine begins with its name in angular brackets and end with an end tag. In this example it begins with [_blockmove]. The end is always [END].

Listing 5 shows the access to a library function in a sample program.

```
Const Bl = 40                ' Defines a block length

Dim Blocklength As Byte
Blocklength = Bl

Dim Block1(bl) As Byte       ' Two blocks of 40 bytes each
Dim Block2(bl) As Byte

Dim I As Byte                ' Index variable

$lib "CK.LIB"                ' Use _blockmove from CK.LIB
$external _blockmove
```

```
Declare Sub Blockmove(source As Byte , Dest As Byte , Byval Length
As Byte)

For I = 1 To Bl                   ' Initialize Block1
  Block1(i) = I * 2
Next
                                  ' Call Blockmove subroutine
Call Blockmove (block1(1) , Block2(1) , Blocklength)

For I = 1 To 40                   ' clear Block1
  Block1(i) = 0
Next

For I = 1 To 40                   ' copy Block2 to Block1 back
  Block1(i) = Block2(i)
Next

End

' Blockmove is the entry for _blockmove assembler routine
Sub Blockmove (source As byte , Dest As byte , Length As Byte)
$asm
  Loadadr Length , Z
  ld _temp2, Z
  Loadadr Source , Z
  Loadadr Dest , X
  rcall _blockmove                ' copy from source to dest
           ' length bytes
$end Asm
Return
```

Listing 5 Copying a memory area (TEST_LIB.BAS)

At the beginning of the program two memory blocks of a length of 40 bytes each are declared. Block1 is (arbitrariy) initialized before the assembler routine `_blockmove` copies block1 to block2.

The BASIC subroutine handles the parameter for the assembler routine only. The copying process takes place exclusively at assembler level.

To compare the runtime of such an assembler routine with a common BASIC subroutine, block1 is cleared for the purpose of copying block2 back to block1 at BASIC level (thereafter).

A runtime measurement is possible in AVR Studio and delivers the following results for the 4 MHz clock frequency:

Routine	Blockmove (...)	For I = 1 To 40 Block1(i) = Block2(i) Next
Runtime	89,6 µs	474,0 µs

2.5.5 Programming Devices

2.5.5.1 AVR

As AVR microcontrollers are in-system programmable (ISP), programming equipment is not required. Rather, the evaluation boards can be used to program and test the first AVR programs.

If evaluation board MCU00100 is used, the AVR ICP910 Programmer needs to be activated. Figure 28 shows the user interface including memory dumps for Flash memory and EEPROM.

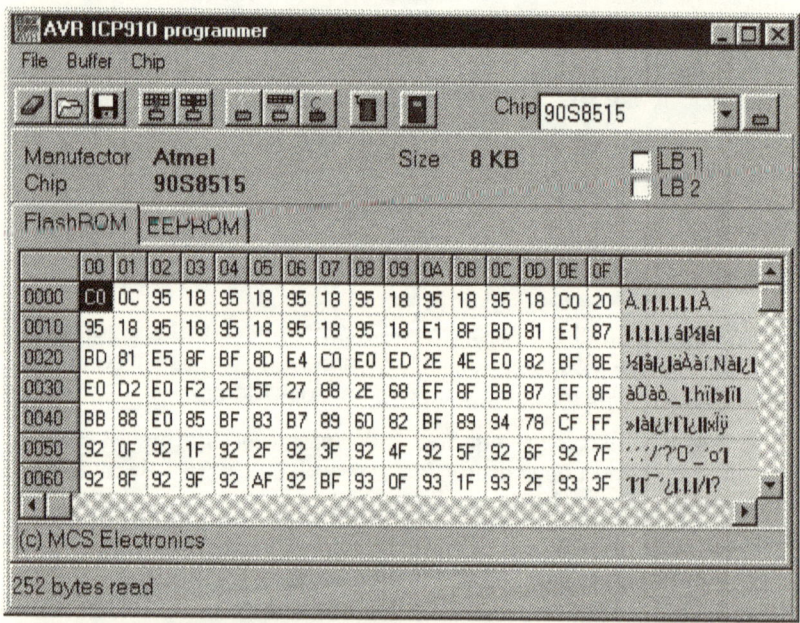

Figure 28   Programming with AVR ICP910

If there is no evaluation board or programmer available, one of the proposals published in the Web are a good choice to consider.

Figure 29 shows the circuit diagram of Jerry Meng's FBPRG Programmer driven from the parallel port of the PC [http://www.qsl.net/fa1fb/]. A lot of people use Jerry's design with success.

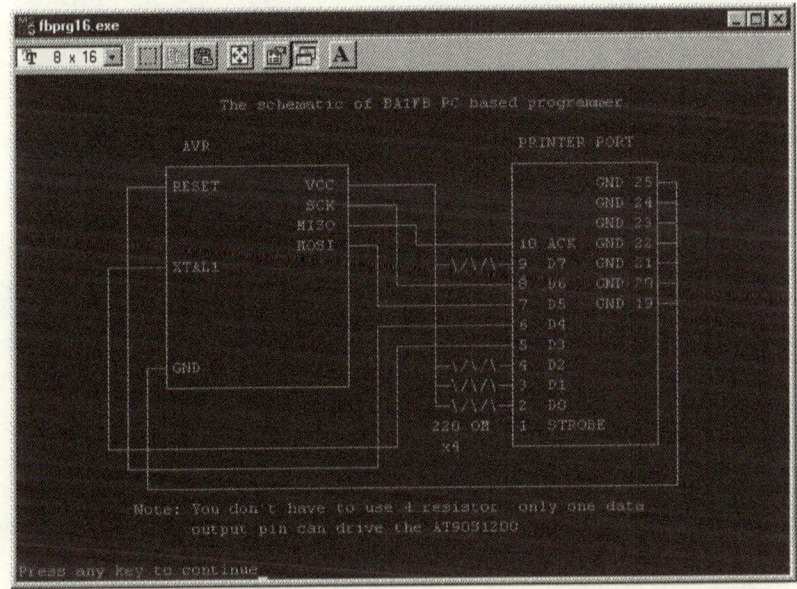

Figure 29   FBPRG Programmer

Figure 30 shows the user interface of the programmer software in a DOS window.

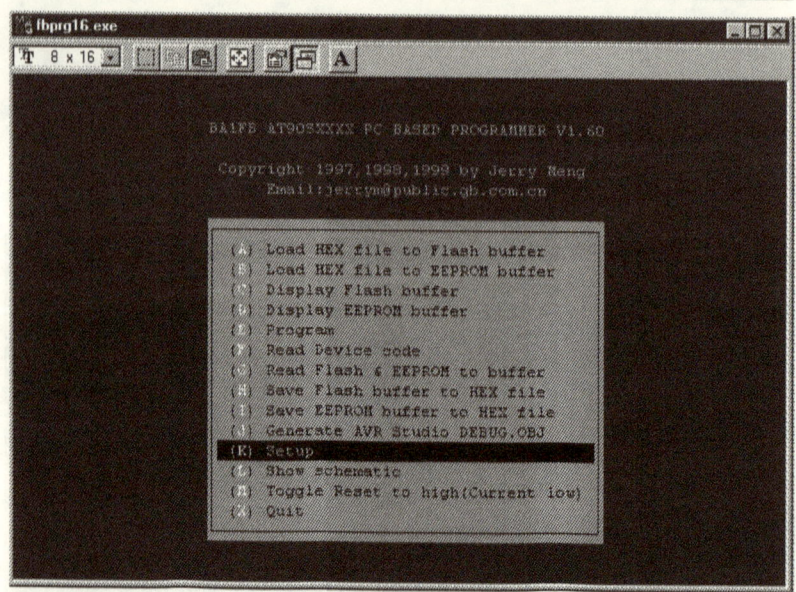

Figure 30   FBPRG in a DOS window

Programmer software and circuit diagram of Jerry Meng's FBPRG Programmer are available for downloading from the author's web site assigned to this book [www.ckuehnel.ch/bascom_buch.htm].

BASCOM-AVR does not support this programmer directly. The programmer software FBPRG.EXE must be linked in menu *Options>Programmer>Other* to BASCOM-AVR.

This is the way to include unknown programmers in both BASCOM IDEs.

2.5.5.2 8051

BASCOM-8051 supports the whole 8051 family of microcontrollers with many memory types and programming needs. It is becessary to choose the right programmer for the microcontroller in use.

The Micro-Pro 51 from Equinox Technologies was used for programming the 8051 derivatives (mostly the AT89C2051) [http://www.equinox-tech.com].

After installing the link to the external programmer, the latter can be run directly from BASCOM-8051. Figure 31 shows the installation of an external programmer.

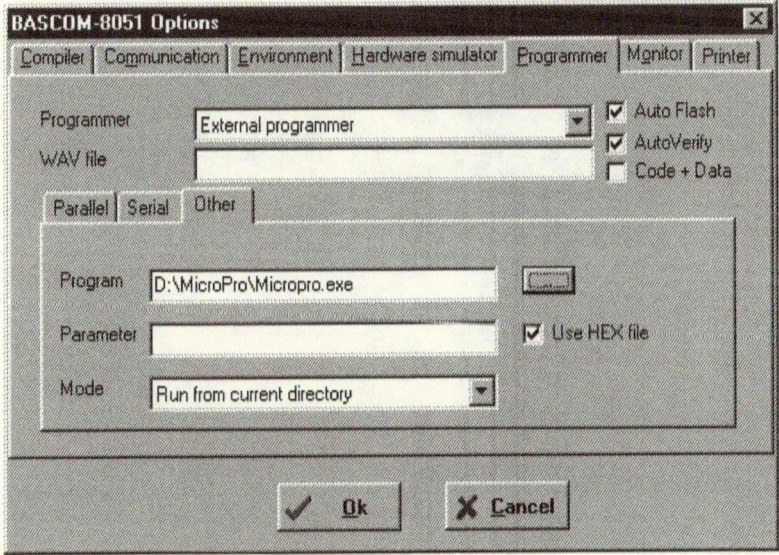

Figure 31   Link to external programmer

This programmer has no special features. Figure 32 shows a loaded hex file in the internal buffer. After programming the result should be comparable with Figure 33.

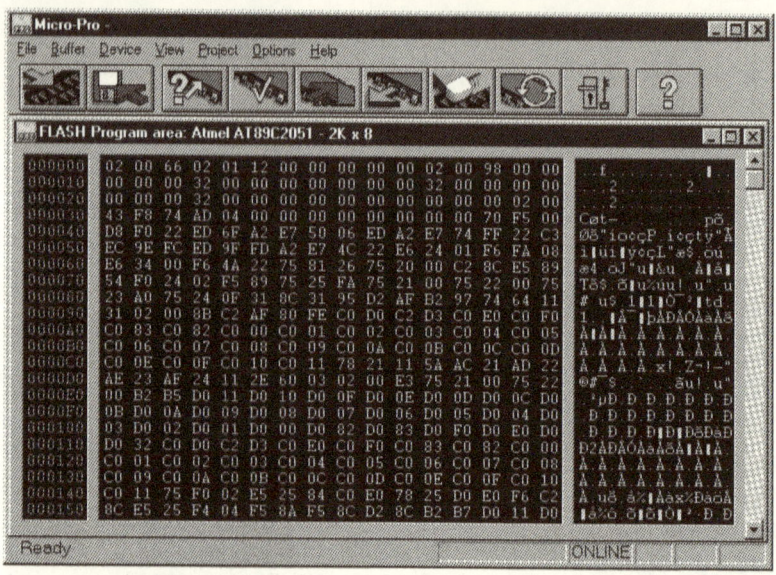

Figure 32 Buffer view

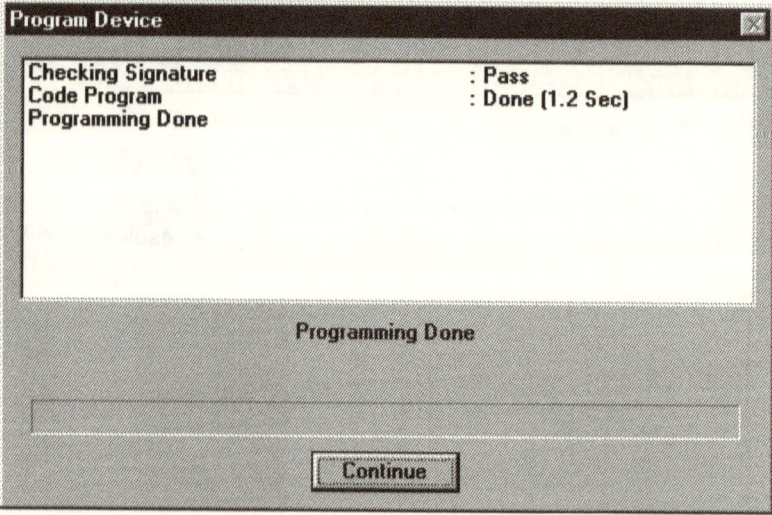

Figure 33 Programming result

After programming the device, the microcontroller must be placed in the target board. For the small AT89C2051 I used the X051 Demo Module from the same manufacturer.

There are many other programming and evaluation devices on the market. Frequently, the manufacturers of microcontrollers offer such devices for first tests or prototyping.

2.6 Hardware for AVR RISC Microcontroller

2.6.1 DT006 AVR Development Board

Dontronics [http://www.dontronics.com] is the producer of the so called SimmSticks. This is a standard that makes use of the well know Simm connectors. There are motherboards and application boards.

For BASCOM-AVR Dontronics designed the DT006. This is a motherboard with integrated Sample Electronics programmer, LEDs, switches and RS-232 serial interface.

So with this PCB you can create a programmer and you can use it as a development board too.

The DT006 board will program the 8, 20, and 28 pin DIP chips on board, and will also program the DT107 (8515 and 4433 footprint), DT104 (2313 footprint) and SIMM100 (8535 footprint) AVR SimmSticks, as well as any AVR target board that has a Kanda type header.

Current burning software is achieved with the programmer software built into Bascom-AVR.

This means, after you have this programmer unit up and running as a development platform, all you need to duplicate the procedure with a stand alone micro, is a single AT90S2313-10-PC micro, and a DT104 PCB and a handful of simple components. Or you can use your own circuit design on a proto board, vero board, your own artwork, whatever.

Figure 34 shows the DT006 AVR Development Board with two Simm expansions slots on the right side (J2, J3). Chapter 5.2 shows the DT006 circuit diagram.

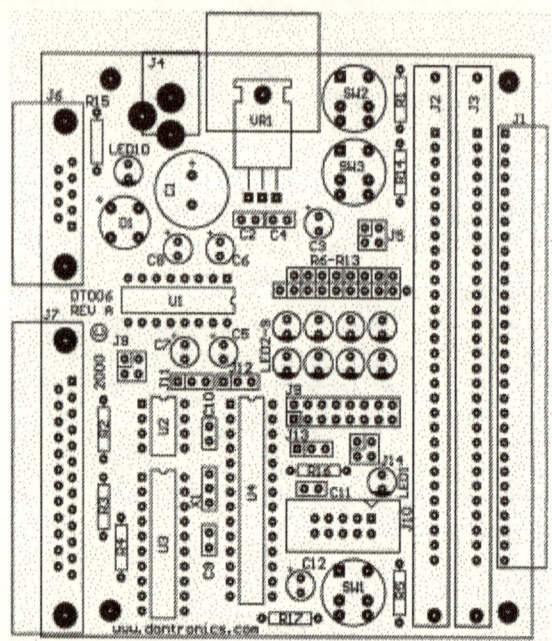

Figure 34 DT006 Board

2.6.2 AVR-ALPHA with AT90S2313

We [http://www.ckuehnel.ch/ask.htm] support starting with the 2313 by the small AVR-ALPHA mini module. Figure 35 shows this module.

All I/O lines are connected to the pins of that module. Prototyping without soldering is possible using a simple breadboard.

All AVR micros are in-circuit programmable and a simple programming adapter fulfills all needs for programming the AVR-ALPHA. This programming adapter can be linked to BASCOM-AVR as an external programmer.

Figure 35 AVR-ALPHA Mini Module

2.7 Instead of "Hello World"

After the introduction of the basic programming procedure as well as the BASCOM Options and Tools, a first and very simple program example will describe the working with BASCOM.

Usually, programs of the "Hello World" class fulfill this exercise. But, the example here is a program controlled by a timer interrupt which I think is a more typical microcontroller program than "Hello World".

Due to the different hardware base of the 8051 and AVR family, the timer example will be explained for both microcontroller families separately.

2.7.1 AVR

Timer0 is an 8-bit timer with a 10-bit prescaler. The timer period can be calculated using the following expression:

$$T = 256 \cdot \frac{prescaler}{f_{OSC}}$$

For a clock frequency of 4 MHz and a prescaler of 1024 a timer period of 0.065536 s is obtained. That means the timer overflows each 0.065536 s and generates an interrupt.

In our program example the assigned interrupt service routine (ISR) increments a byte variable and toggles an I/O pin. Listing 6 shows the source text of program SIM_TIMER.BAS.

```
' SIM_TIMER.BAS for AVR

Dim A As Byte                   ' Temporary Variable

Ddrb = 255                      ' PortB is output
Portb = 255                     ' All outputs Hi

' Configure the timer to use the clock divided by 1024
Config Timer0 = Timer , Prescale = 1024

On Timer0 Timer0_isr            ' Jump to Timer0 ISR

Enable Timer0                   ' Enable the timer interrupt
Enable Interrupts               ' Enable Global Interrupt

Do
' Do nothing
Loop

Timer0_isr:                     ' Interrupt Service Routine
    Incr A                      ' Increment Variable A
    Portb.0 = Not Portb.0       ' Toggle Portb.0
Return
```

Listing 6 Timer program for AVR (SIM_TIMER.BAS)

2.7.2 8051

Timer0 operates in Mode 2 as a 16-bit timer. The timer period can be calculated using the following expression:

$$T = 65536 \cdot \frac{12}{f_{OSC}}$$

For a clock frequency of 11.059 MHz and a fixed prescaler of 12, a timer period of 0.07111 s is obtained. That means the timer overflows each 71 ms and generates an interrupt.

In our program example the assigned interrupt service routine (ISR) increments a byte variable and toggles an I/O pin. Listing 7 shows the source text of program SIM_TIMER.BAS.

```
' SIM_TIMER.BAS for AT89C2051

Dim A As Byte                    ' Temporary Variable

P1 = 255                         ' All outputs Hi

' Configure timer0 as 16-bit timer
Config Timer0 = Timer , Mode = 1
Start Timer0

On Timer0 Timer0_isr             ' Jump to Timer0 ISR

Enable Timer0                    ' Enable the timer interrupt
Enable Interrupts                ' Enable Global Interrupt

Do
' Do nothing
Loop

Timer0_isr:                      ' Interrupt Service Routine
    Incr A                       ' Increment Variable A
    P1.0 = Not P1.0              ' Toggle P1.0
Return
```

Listing 7 Timer program for 8051 (SIM_TIMER.BAS)

2.7.3 Things in Common

When comparing Listing 6 with Listing 7, only a few differences can be seen to exist; the major part does not differ from eachother.

At first, a variable A is declared as byte. The format (here byte) defines the memory allocation to the variable.

The timer overflow interrupt toggles an I/O pin. For AVR we use Pin0 of PortB (PortB.0) and for 8051 Pin0 of Port1 (P1.0).

For the AVR, a data direction register initializes a pin as input or output. Therefore, at least the pin toggled must be an output. To simplify matters all pins of PortB are declared as outputs (DDRB = 255) and set to Hi afterwards (PORTB = 255). For 8051, all Pins of Port1 are set to Hi (P1 = 255) only.

The timer configurations of the microcontroller family differ from each other; see the description in the previous two chapters.

Finally, the interrupts must be enabled. Enable Timer0 enables the timer interrupt and Enable Interrupts enables the global interrupts in the last initialization step.

Following this initialization the program enters its main loop (`Do..Loop`) where nothing is to be done in this example.

The declaration of an interrupt service (ISR) routine in BASCOM is performed in the same way as the declaration of a normal subroutine. The compiler replaces `Return` by the required `Reti` (Return from Interrupt) and supports the Push and Pop of all registers.

Inside the ISR `Timer0_isr` variable A is incremented (`Incr A`) and Pin0 of PortB (AVR) and Port1 (8051), respectively, will be toggled afterwards. That means reading Pin0, inverting the value and writing back (`Portb.0 = Not Portb.0` and `P1.0 = Not P1.0`, respectively).

Next, input this program or open it after downloading from our web site. Compiling and debugging is explained for BASCOM-AVR only but do not differ for BASCOM-8051.

Figure 36 shows the source text of program SIM_TIMER.BAS opened in the BASCOM-AVR Editor.

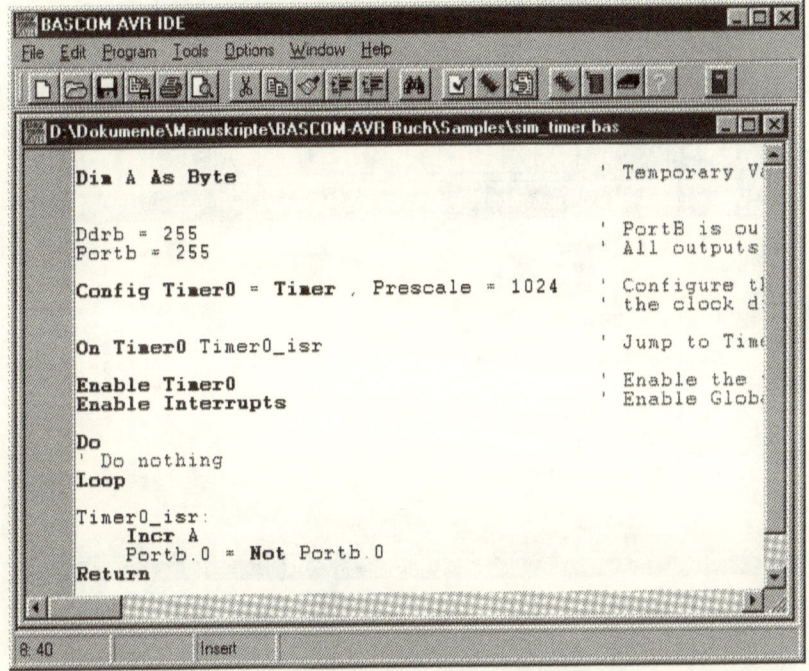

Figure 36 Source text in BASCOM-AVR Editor

Before the first compilation the options must be set. The parameters for the serial interfaces (I^2C, SPI and 1-wire) and LCD are not relevant here and can be set as desired.

Before a complete compilation, it may help to check the syntax. Start the syntax check from menu *Program>Syntax Check* or *Ctrl+F7*.

Figure 37 shows a syntax check with errors. By double-clicking the error line the last "e" is seen to be missing in instruction Enable.

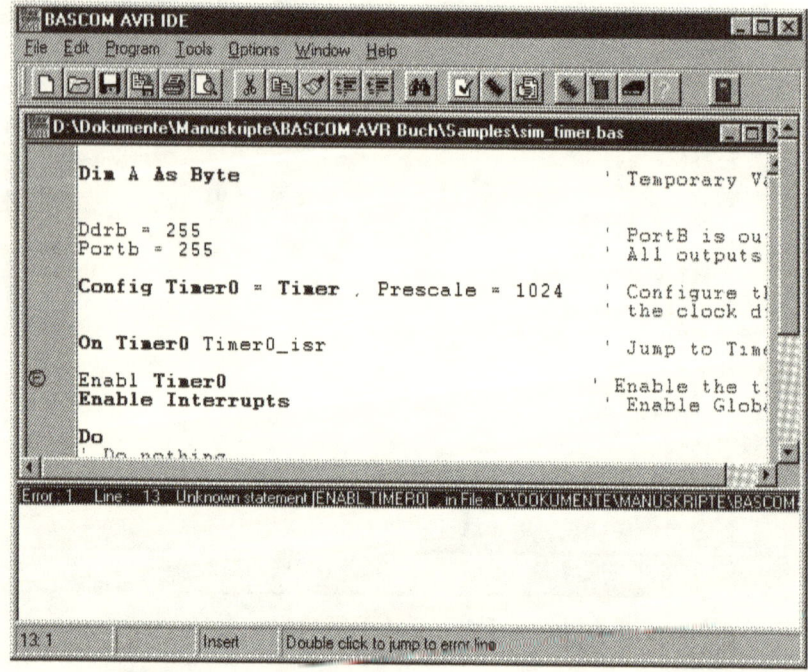

Figure 37   Result of syntax check

When the missing character is entered, the syntax check will show no error anymore, and the compilation will be faultless as well. Start the compilation from menu **Program>Compile** or **F7**.

As expected there is no error after compilation. Look for the result by clicking **Program>Show Result** or **Ctrl+W**. Listing 8 shows the report file SIM_TIMER.RPT generated for BASCOM-AVR.

```
Report      : SIM TIMER
Date        : 10-31-1999
Time        : 19:07:06

Compiler    : BASCOM-AVR LIBRARY V 1.05, Standard Edition
Processor   : 90S8515
SRAM        : 200 hex
EEPROM      : 200 hex
ROMSIZE     : 2000 hex

ROMIMAGE    : FA hex  -> Will fit into ROM
BAUD        : 9600 Baud
XTAL        : 4000000 Hz
BAUD error  :   0.16%

Stackstart   : 25F hex
S-Stacksize  : 20 hex
S-Stackstart : 240 hex
Framesize    : 32 hex
Framestart   : 20D hex

LCD DB7     : PORTB.7
LCD DB6     : PORTB.6
LCD DB5     : PORTB.5
LCD DB4     : PORTB.4
LCD E       : PORTB.3
LCD RS      : PORTB.2
LCD mode    : 4 bit
```

Variable	Type	Address(hex)	Address(dec)
COUNTER0		0032	50
TIMER0		0032	50
COUNTER1	Word	004C	76
TIMER1	Word	004C	76
CAPTURE1	Word	0044	68
COMPARE1A	Word	004A	74
COMPARE1B	Word	0048	72
PWM1A	Word	004A	74
PWM1B	Word	0048	72
ERR		0006	6
A	Byte	0060	96

Warnings:

Listing 8   Report file for BASCOM-AVR (SIM_TIMER.RPT)

Listing 9 shows the report file generated for BASCOM-8051.

```
Compiler    : BASCOM 8051 LIBRARY V 2.04
Processor   : AT89C2051
Report      : SIM_TIMER
Date        : 12-27-2000
Time        : 15:20:03

Baud Timer  : 1
Baudrate    : 0
Frequency   : 11059200
ROM start   : &H0
RAM start   : &H0
LCD mode    : 4-bit
StackStart  : &H22
Used ROM    : &HAD         173 (dec)    > Ok

------------------------------------------------------------
Variable                       Type    Address(hex)  Address(dec)
------------------------------------------------------------
ERR                            Bit       0004          4
A                              Byte      0021          33
CONSTANTS
------------------------------------------------------------
Constant                       Value
------------------------------------------------------------
```

Listing 9   Report file SIM_TIMER.RPT for BASCOM-8051

2.7.4 Simulation

In the next step, the simulator can be started from menu **Program>Simulate** or by pressing **F2**.

The simulator of the BASCOM IDE has been referred to already. So let's use here the simulator of the AVR Studio for the BASCOM-AVR example.

Caution: If you have to go deep into the compiled code, using the AVR Studio has some advantages. When a functional simulation is sufficient, using the internal simulator will be simpler.

Load the generated Obj-File from menu **File>Open** or press **Ctrl+O**. Figure 38 shows the simulator with three open windows.

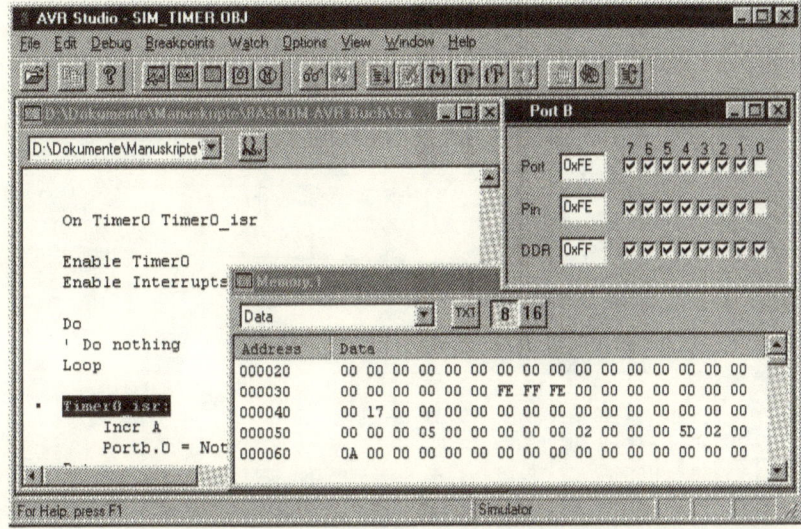

Figure 38 Simulation von SIM_TIMER.BAS in AVR Studio

The source text can be seen on the left side. We placed a break point to the ISR. The top right window shows all bits of PortB. The bottom right window shows a memory dump of data memory. These windows can be opened from menu *View>Peripheral>Port>PortB* or *View>New Memory View*.

Simulation can start when the AT90S8515 is selected in the simulation option (*Options>Simulation Options*).

Start the simulation from menu *Debug>Go* or *F5*. The simulation stops at the break point. All changes at Pin0 of PortB and in memory location 60_H are visible. The timer period in simulation depends on the fastness of the PC used. With the author's PC a timer period of about five seconds was achieved.

However, a simulation is not all in life. Therefore the program is burnt into the microcontroller and the program checked in the target hardware.

Evaluation board MCU00100 must be connected to COM1 of the PC before starting the programmer from menu *Program>Send to Chip* or *F4*. Figure 39 shows the user interface of the programmer.

The AT90S8515 used in evaluation board MCU00100 has already

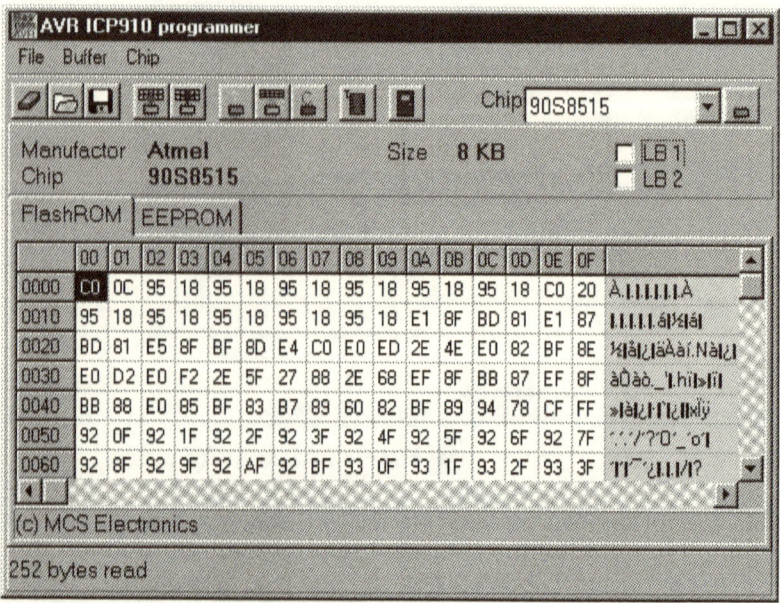

Figure 39   BASCOM-AVR ICP910 Programmer

been identified already. The generated code is visible in FlashROM. Load the program into the flash memory of AT90S8515 from menu *Chip>Autoprogram*.

Immediately after Verify, the program starts and the LED connected to Pin0 of PortB blinks at the programmed rate.

This first example should explain the fundamental project work with BASCOM. It should be clear that planning the resources like the allocation of I/O pins and so on is independent of the used programming language or environment. This step must be finished before coding. Later, any resulting collisions can be repaired at greater expenses only.

2.8 BASCOM Help System

If you need help for any BASCOM instruction, place the cursor to the respective key word and press function key F1. Figure 40 shows the opening help window with explanations.

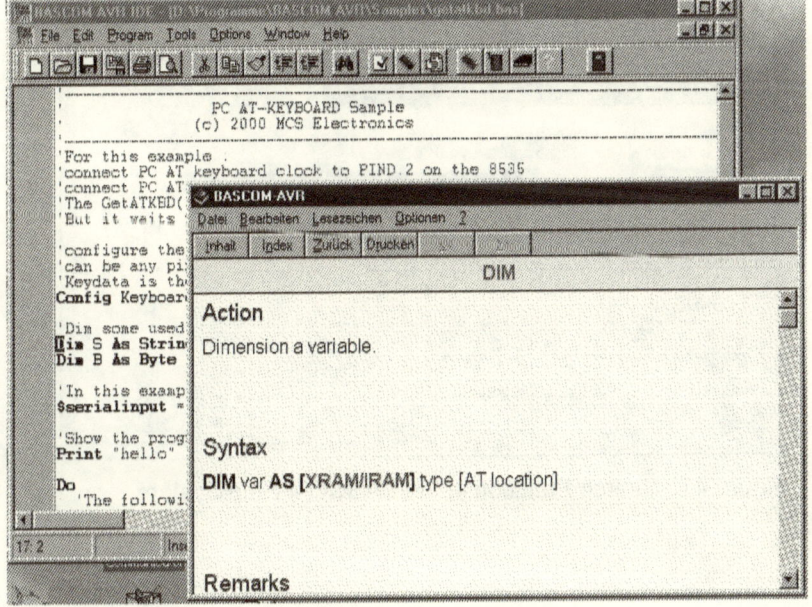

Figure 40   Press F1 for Help

Just as important as the explaining text are additional program examples which describe the use of instructions and/or functions.

Furthermore, the help system has a very comfortable index and search system. Figure 41 shows a search for "interrupt" information and resulting hints.

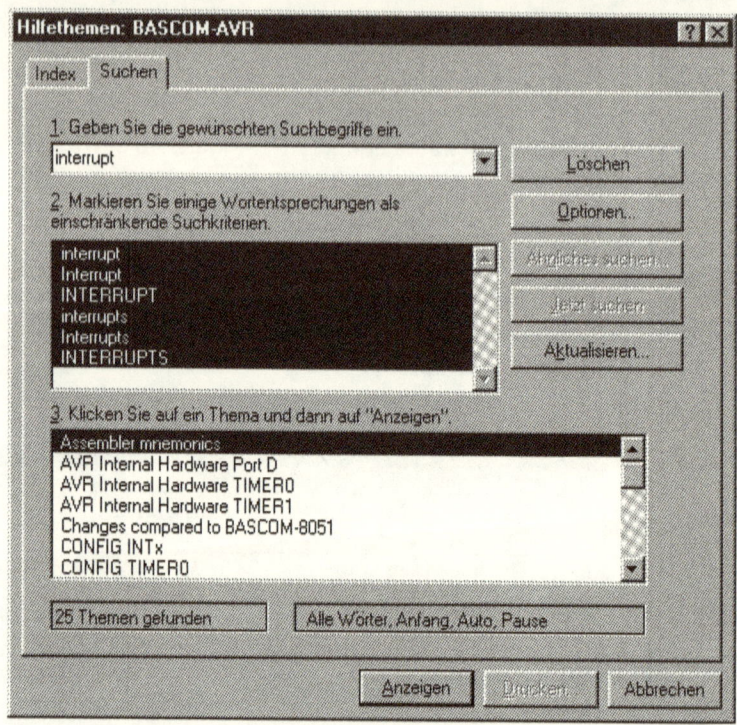

Figure 41 Search Function in the BASCOM Help System

3 Some BASCOM Internals

This chapter describes some BASCOM details which caused some responses and queries in the past.

Caution: Since BASCOM has a very powerful help system, there is no list of compiler directives and instructions in this book.

Please use the Help System first. A lot of newsgroup queries can be answered this way.

3.1 Building new instructions

BASCOM's subroutine construct is a powerful means for generating new instructions. A simple example will demonstrate it.

In the example we generate an instruction that toggles some pins of port P1 of an 8051 microcontroller. The instruction shall have one parameter - the toggle mask.

We define subroutine Toggle_p1(x) that reads, masks and writes back port P1 (P1 = P1 Xor X). If this subroutine is declared at the beginning of the program, there are two ways for calling it at a later time.

Call Toggle_p1(mask) and Toggle_p1 mask are two equivalent subroutine calls. The second kind of call is marked bold in the program example. It looks like a new instruction.

```
Dim Mask As Byte , X As Byte
Mask = &B11000011              ' Toggle mask

' Declaration of instruction toggle_p1
Declare Sub Toggle_p1(x As Byte)

Do
    Call Toggle_p1(mask)       ' Subroutine call
    Toggle_p1 Mask             ' Usage of new instruction
Loop

End

' Defintion of subroutine
Sub Toggle_p1(x As Byte)
    P1 = P1 Xor X
End Sub
```

We can go into the simulator and see the equality again.

In the single-step mode, we set port P1 to &HA5, for example, and see P1 toggling from &HA5 to &H66 and vice versa.

```
P1      &HA5      1010 0101
Mask              1100 0011
P1      &H66      0110 0110
```

Figure 42 shows the simulator window with port P1 toggled.

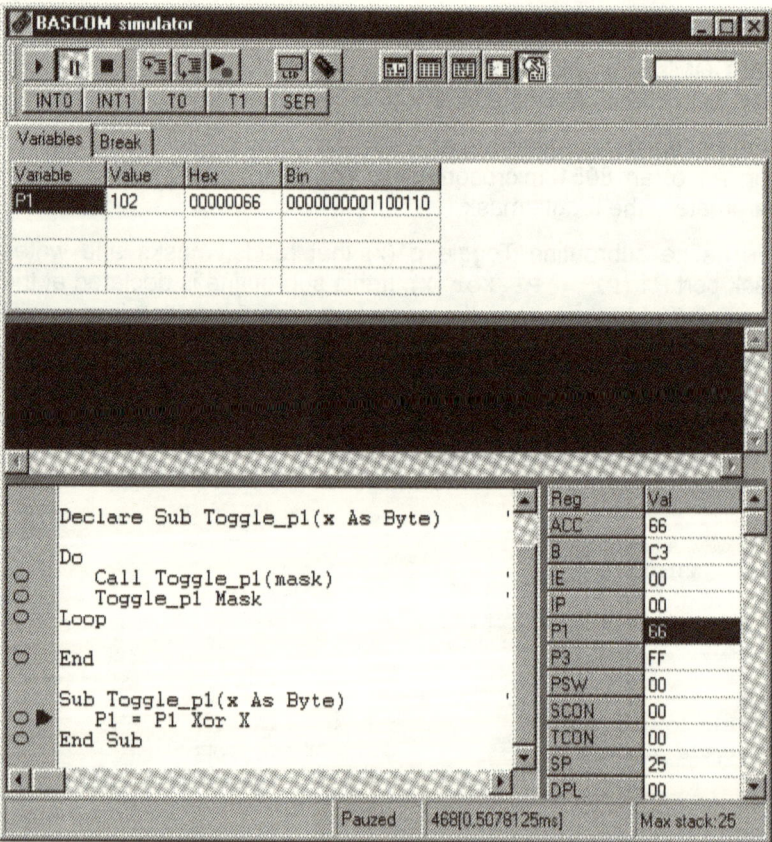

Figure 42   Subroutine call in simulator of BASCOM-8051

There are minor differences between BASCOM-8051 and BASCOM-AVR as regards the declaration of subroutines. The next chapter describes in detail the parameter passing by reference or by value in BASCOM-AVR only.

3.2 Parameters for Subroutines in BASCOM-AVR

A wrong parameter handling BYREF or BYVAL is a frequent reason for errors in application programs. A simple example will give more clarity.

In the next example, a mask function cutting the high nibble of a value multiplied by four is defined.

Value A and mask B are parameters of a function to be defined. The result is saved in variable Z. The binary representation of this task is as follows:

```
A                   10101010    for example
Shift A , Left , 2  10101000    2*A
B                   00001111    mask
Z                   00001000    result
```

The program reads as follows:

```
' Subroutine in BASCOM-AVR

Dim X As Byte , Y As Byte , Z As Byte

X = &B10101010
Y = &B00001111

Declare Function Mask (byval a As Byte , B As Byte) As Byte

Z = Mask(x , Y)

End

Function Mask (byval A As Byte , B As Byte) As Byte
    Shift A , Left , 2
    Mask = A And B
End Function
```

Running the program in the single-step mode reveals that variable x is unchanged after access to the function. Figure 43 shows the unchanged variable x after access to function mask().

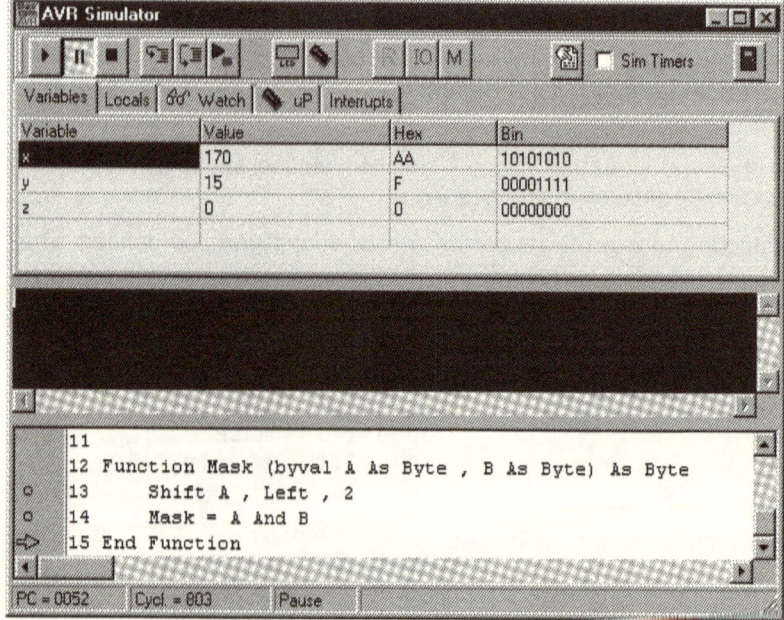

Figure 43   Parameter passing BYVAL

When cutting keyword BYVAL, the default parameter passing BYREF is active. In this case the variable changes from &HAA to &HA8 after access to the function mask (Shift A , Left , 2); see Figure 44.

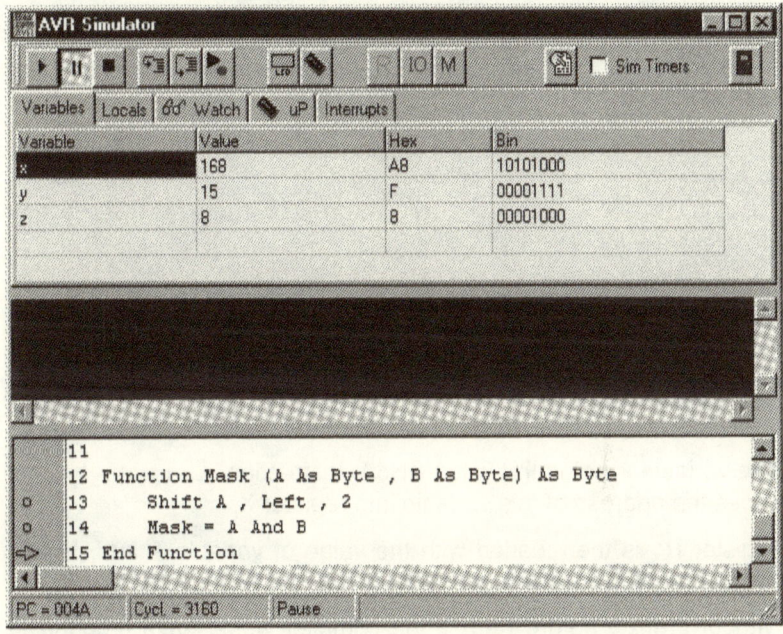

Figure 44 Parameter passing BYREF (default)

3.3 BASIC & Assembler

In some cases a direct effect on the code is needed. If a desired function is not in the instruction set, it can be defined as a BASIC or Assembler subroutine.

Of importance is that BASCOM supports the mixing of BASIC and Assembler.

The compiler recognizes most assembler mnemonics automatically. Exceptions are SWAP and, additionally, SUB and OUT for BASCOM-AVR. These mnemonics are reserved words of BASIC and therefore have a higher priority as the assembler mnemonics.

However, using prefix ! makes the compiler recognize that word as assembler mnemonics, too.

Short examples for both microcontroller families demonstrate the use of the assembler in a BASIC source file.

3.3.1 AVR

The assembler is based on the standard AVR mnemonics.

In the following program examples, the assembler instructions are marked in bold.

```
Dim A As Byte              ' Bytevariable
A = &H5A                   ' Initialize Variable

Loadadr A , X              ' Load Address of A into X

Ld R1, X                   ' Load R1 with contents where
                           ' X is pointing to

!SWAP R1                   ' Swap nibbles
```

Byte variable A holds the value &H5A. Instruction `Loadadr A , X` places the address of this variable into register X.

Register R1 is then loaded with the value of variable A and, finally, the content of register R1 is swapped.

Without prefix ! before `swap`, the compiler would have recognized `swap` as a BASIC instruction.

Another possibility is the use of the compiler directives `$asm` and `$asm end`. Normal assembler mnemonics can be placed between these two directives.

```
Dim A As Byte              ' Bytevariable
A = &H5A                   ' Initialize Variable

Loadadr A , X              ' Load Address of A into X
$asm
    Ld R1, X               ' Load R1 with contents where
                           ' X is pointing to

    Swap R1                ' Swap nibbles
$end Asm
```

Run these examples in the simulator to see how such includes work.

It is a matter of taste what kind of notation one prefers. Functionally, both examples are equivalent.

Take care when manipulating registers directly! BASCOM-AVR uses some registers. R4/R5 serve as a pointer to the stack frame. R8/R9 serve as data pointer for the READ instruction. R6 contains a few bit variables:

R6.0   Flag for integer-word conversion
R6.1   Temporary bit for bit swap
R6.2   Error bit (ERR)
R6.3   Show/Noshow bit of INPUT instruction

Caution: Do not change these registers in any assembler included.

Other registers will be used independence of the BASIC instruction referred to.

3.3.2 8051

The assembler is based on the standard Intel mnemonics.

In the following program examples, the assembler instructions are marked in bold.

```
Dim A As Byte              ' Bytevariable
A = &H5A                   ' Initialize Variable

Placeadres A , R0          ' Load R0 with address
                           ' from variable A

MOV A,@R0                  ' Load ACC with contents
                           ' of variable A

!SWAP A                    ' Swap nibbles
```

Byte variable A holds the value &H5A. Instruction `Placeadres A , R0` places the address of this variable into register R0.

The accumulator is then loaded indirectly with the value of variable A and, finally, the content of the accumulator is swapped.

Without prefix ! before `swap`, the compiler would have recognized `swap` as a BASIC instruction.

Another possibility is the use of compiler directives $asm and $asm end. Normal assembler mnemonics can be placed between these two directives.

```
Dim A As Byte              ' Bytevariable
A = &H5A                   ' Initialize Variable

Placeadres A , R0          ' Load R0 with address
                           ' from variable A

$asm
    MOV A,@R0              ' Load ACC with contents
                           ' of variable A
    Swap A                 ' Swap nibbles
$end Asm
```

A third way simplifies access to the variable by a different notation.

```
Dim A As Byte              ' Bytevariable
A = &H5A                   ' Initialize Variable

$asm
    MOV A,{A}              ' Load ACC with contents
                           ' of variable A
    Swap A                 ' Swap nibbles
$end Asm
```

Run these examples in the simulator to see how such assembler includes work.

It is a matter of taste what kind of notation one prefers. Functionally, all three examples are equivalent.

Caution: Take care when directly manipulating registers! BASCOM-8051 uses the registers ACC, B and SP. Do not change these registers in any assembler included.

4 Applications

This chapter describes the applications both microcontroller families can be used for. It is the underlying hardware of the microcontroller concerned that is responsible for any differences in the programs.

The program examples were first set up with BASCOM-AVR. Hints for porting the AVR examples to 8051 are included. In some cases we discuss the solutions which are dependent on the microcontroller used. In other cases the differences are insignificant.

4.1 Programmable Logic

Logical devices query input lines (input pattern) and assign a defined bit pattern to the output lines. The logical relations can be expressed by way of a table. In the example the following relations are intended to be implemented:

A7	A6	A5	A4	A3	A2	A1	A0	Q7	Q6	Q5	Q4	Q3	Q2	Q1	Q0
1	1	1	1	1	1	1	0	1	1	1	1	0	0	0	0
1	1	1	1	1	1	0	1	0	0	0	0	1	1	1	1
x	x	x	x	x	x	1	1	1	1	1	1	1	1	1	1

The logical devices should have eight inputs A7..A0 and eight outputs Q7..Q0. Each bit pattern at an input has a corresponding bit pattern at an output.

For eight input lines we obtain 256 different bit patterns, and the table would be very long.

In the table we use, there are only three different bit patterns at the output. Therefore, the table poses no problem.

Interpreting the table, we find the following results:

- If A = &B11111110 then set Q = &B11110000.
- If A = &B11111101 then set Q = &B00001111.
- In all other cases set Q=&B11111111.

Shown in Fig. 43 is the whole circuit including clock generation and reset circuitry for an application using an AVR microcontroller.

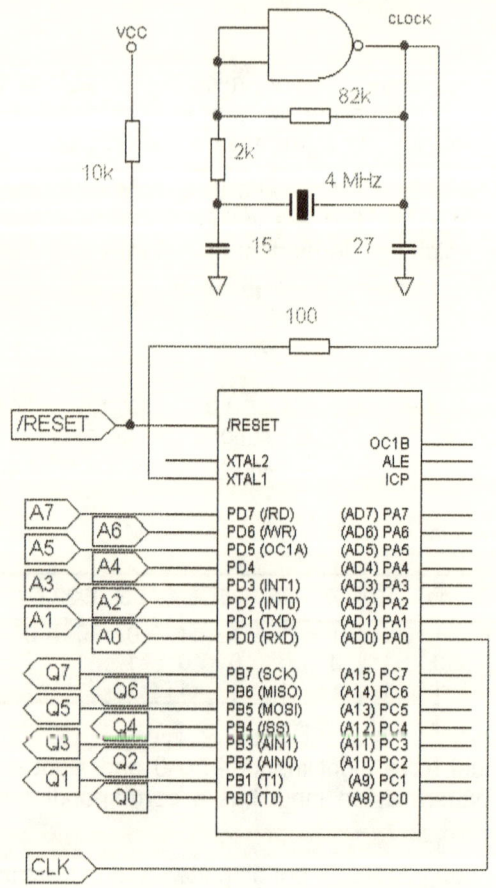

Figure 45   AT90S8515 as a logical device

The clock and reset components are always the same and will be omitted in the next circuit diagrams. Supply voltage and ground are normally not drawn, either.

The eight input lines A7..A0 go to PortD. PortB drives the eight output lines. Keys on the evaluation board used are connected to PortD. PortB is connected to LEDs with resistors in series.

Program LOGIC.BAS waits for a pulse (rising edge followed by a falling edge) at input CLK and, thereafter reads the input lines at

PortD. In a case structure the bit pattern is evaluated and the result forces the pins on PortB.

The BITWAIT instructions query Pin0 of PortA and block the program until the mentioned pulse is detected.

The bit pattern of the input is saved in variable A. Variable Q contains the bit pattern of the output.

The pins have an internal pull-up resistor, which is activated by setting the port line. PORTD.x = 1 activates the pull-up resistor on the respective I/O line. In this program example, the whole port will be set (Listing 10).

```
' Logic with AT90S8515

Dim A As Byte , Q As Byte

Config Porta = Input
Porta = 255                      ' Pull-up active

Config Portb = Output

Config Portd = Input
Portd = 255                      ' Pull-up active

Do
    bitwait Pina.0 , Set
    bitwait Pina.0 , Reset
    A = Pind
    Select Case A
        Case &B11111110 : Q = &B11110000
        Case &B11111101 : Q = &B00001111
        Case Else Q = &B11111111
    End Select
    Portb = Q
Loop

End
```

Listing 10   Logical device with AT90S8515 (LOGIC.BAS)

Input CLK triggers the data input of the input lines at PortD. So as to get periodic queries of the input lines, a timer can be used for triggering. The circuit remains unchanged. Input CLK has no function now.

Timer applications will be discussed in the next chapter.

Listing 11 shows the timer controlled logic device. The complete I/O handling is here accommodated in the interrupt handler.

```
' Logic with AT90S8515

Dim A As Byte , Q As Byte

Config Portb = Output
Portb = 255                     ' All outputs Hi

Config Portd = Input

' Configure the timer to use the clock divided by 1024
Config Timer0 = Timer , Prescale = 1024

On Timer0 Timer0_isr            ' Jump to Timer0 ISR

Enable Timer0                   ' Enable the timer interrupt
Enable Interrupts               ' Enable Global Interrupt

Do
   Nop
Loop

End

Timer0_isr:
   A = Pind
   Select Case A
      Case &B11111110 : Q = &B11110000
      Case &B11111101 : Q = &B00001111
      Case Else Q = &B11111111
   End Select
   Portb = Q
Return
```

Listing 11 Timer controlled logic devices (LOGIC1.BAS)

When deciding to use an 8051 microcontroller a device with enough I/O lines is required. The next program examples are based on the AT89S8252. Listing 12 shows the slightly modified program. A comparison with Listing 10 shows modifications only for port I/O.

P1 is the input and P3 the output for the logical signals. P2.0 serves as clock input here.

```
' Logic with AT89S8252

Dim A As Byte , Q As Byte

P1 = 255                        ' Pull-up active
P2 = 255                        ' P2.0 is CLK input

Do
    Bitwait P2.0 , Set
    Bitwait P2.0 , Reset
    A = P1                      ' Read P1
    Select Case A
        Case &B11111110 : Q = &B11110000
        Case &B11111101 : Q = &B00001111
        Case Else Q = &B11111111
    End Select
    P3 = Q      ' Write P3
Loop

End
```

Listing 12   Logical device with AT89S8252 (LOGIC.BAS)

4.2 Timer and Counter

As the timers/counters of the 8051 and AVR microcontrollers differ from each other, the timers will be described separately.

Timer and counter denote different modes of the same hardware. To simplify description, the term timer will be used in all general explanations.

Caution: Please read the documentation of the manufacturer very carefully. The correct setup of some registers is the key to a correct implementation of the required functions. In case of a wrong setup debugging can be very difficult.

4.2.1 AVR

The AVR microcontrollers have different internal timers. The 8-bit timer has already been used for simple timer functions.

Since the 16-bit timer offers far more flexibility than the 8-bit timer, it will be primarily dealt with here.

Caution: The pinout for the alternative functions such as clock inputs T0 and T1, differs for the various types of the AVR family.

All timer program examples given below refer to the AT90S8515.

4.2.1.1 Timer

Timer0 is an 8-bit timer and Timer1 a 16-bit timer. Each timer has a 10-bit prescaler. The maximum timer period can be calculated using the following equation:

$$T = 2^N \cdot \frac{prescaler}{f_{OSC}}$$

N = 8 for Timer0 and N = 16 for Timer1. The prescaler may have a value of 1, 8, 64, 256 or 1024. The next tables show the resolution and maximum timer period for Timer0 and Timer1 for a clock frequency of 4 MHz.

Timing for Timer0 at 4 MHz					
Prescaler	1	8	64	256	1024
max. Timer Period in ms	0,064	0,512	4,096	16,384	65,536
Resolution in ms	0,00025	0,002	0,016	0,064	0,256

Timing for Timer1 at 4 MHz					
Prescaler	1	8	64	256	1024
max. Timer Period in s	0,016	0,131	1,049	4,194	16,777
Resolution in s	0,00025	0,002	0,016	0,064	0,256

The next example is a clock generator blinking an LED once per second.

The maximum period for Timer1 with a prescaler of 64 is 1.049 seconds. To get the period of one second exactly we have to shorten this time by 49 ms. The Output Compare Mode of Timer1 can reduce the timer period.

Derived from the equation above we find the following formula

$$OutputCompare = \frac{f_{OSC}}{prescaler} \cdot T_{Soll}$$

and with the known parameters we get an output compare value of 62500. This value must be saved in the Output Compare Register.

Listing 13 shows the configuration of Timer1 as timer with a prescaler of 64.

```
Dim New_time As Byte
Dim Temp As Byte
Dim Seconds As Byte
Dim Minutes As Byte
Dim Hours As Byte
Dim Key As Byte

Const True = 1
Const Reload = 62500

Config Timer1 = Timer , Prescale = 64
Ocr1ah = High(reload)
Ocr1al = Low(reload)              ' Reload Timer1 for
                                  ' Period of 1 sec
Tccr1a = 0                        ' Disconnect OC1A from T/C1
Set Tccr1b.3                      ' Reset T/C1 after Compare

Config Portb = Output
Portb = 255   ' All outputs Hi

On Compare1a Timer1_isr           ' Jump to Timer1 ISR

Enable Compare1a                  ' Enable the timer interrupt
Enable Interrupts                 ' Enable Global Interrupt

Do
    Key = Pind
    If Key = &H7F Then
        Seconds = 0
        Minutes = 0
        Hours = 0
    End If
    While New_time = True
        If Seconds = 60 Then
            Seconds = 0 : Incr Minutes
        End If
        If Minutes = 60 Then
            Minutes = 0 : Incr Hours
        End If
        If Hours = 24 Then Hours = 0
        Temp = Makebcd(seconds)
        If Key = &HFE Then Temp = Makebcd(minutes)
        If Key = &HFD Then Temp = Makebcd(hours)
        Portb = Not Temp
        New_time = Not True
    Wend
Loop

End
```

```
Timer1_isr:
    New_time = True
    Incr Seconds
Return
```

Listing 13   Second-Timer with Timer1 (TIMER3.BAS)

Timer1 operates as an up-counter. When the timer count is equal to the content of Output Compare RegisterA, a compare interrupt occurs. To start a new timer period, bit CTC1 of control register TCCR1B must be set.

To avoid unintentional changes in timer control registers TCCR1A and/or TCCR1B, instruction `CONFIG TIMER1...` should be used at program start before any other timer configurations.

From instruction

```
Config Timer1 = Timer , Prescale = 64
```

BASCOM-AVR generates the following assembler code:

```
LDI     R24,0x00        ; 0x00 = 0b00000000 = 0
OUT     0x2F,R24
LDI     R24,0x03        ; 0x03 = 0b00000011 = 3
OUT     0x2E,R24
```

Register TCCR1A is reset to &H00. Outputs OC1A and OC1B are disconnected from Timer1. PWM is deactivated. Register TCC1B is set to &H03 switching the prescaler 64.

The CTC1 bit in register TCCR1B must be set separately. Instruction `Set Tccr1b.3` can do the job without exerting any influence on other bits in TCCR1B.

From instruction

```
Set Tccr1b.3
```

BASCOM-AVR generates the following assembler code:

```
IN      R24,0x2E
ORI     R24,0x08        ; 0x08 = 0b00001000 = 8
OUT     0x2E,R24
```

After enabling the compare interrupt, the program enters an endless loop showing the time in the BCD format. Seconds, minutes or hours can be displayed by striking the respective keys.

The Output Compare Function of Timer1 generates a compare interrupt when the timer is equal to the compare value. The interrupt handler sets flag `New_time` and increments variable `Seconds`. A reload of Timer1 is not required because Timer1 is reset on compare event.

Timer0 is less comfortably equipped, and reloading must be implemented in the software. The procedure is demonstrated with a 50 ms timer.

At a clock of 4 MHz and a prescaler of 1024, it will take 195 cycles to get a timer period of 50 ms.

Timer0 has the overflow interrupt available only. Timer0 must be loaded with a value of 256 - 195 to get an overflow after 195 cycles.

Listing 14 shows the initialization of Timer0 and PortB and an endless loop as the main program.

On Timer0 overflow the instruction `Load Timer0 , Reload` reloads the timer. Calculation 256 - Reload is performed internally.

```
Const Reload = 195            ' Reload value for Period of 50 ms

Config Timer0 = Timer , Prescale = 1024

On Timer0 Timer0_isr          ' Jump to Timer1 ISR

Config Portb = Output

Enable Timer0                 ' Enable the timer interrupt
Enable Interrupts             ' Enable Global Interrupt

Do
    Nop
Loop

End

Timer0_isr:
      ' Reload Timer0 for Period of 50 ms
      Load Timer0 , Reload
      Portb.0 = Not Pinb.0    ' Toggle Portb.Pin0
Return
```

Listing 14   Clock generation using Timer0 (TIMER0.BAS)

Especially when manipulating the internal registers it is recommended to inspect the initialized registers or the generated code with the simulator.

As shown in the following assembler list, all internal registers and the status register are saved on stack at the beginning of every interrupt service routine (ISR). Only after this pushing will the activities of the ISR start.

In the example, the first activity (marked gray) is loading register TCNT0 with the value of 256-195 = &H3D.

Instruction		Cycles	TCNT0 (Prescaler=1)
PUSH	R0	2	&H06
PUSH	R1	2	&H08
PUSH	R2	2	&H0A
PUSH	R3	2	&H0C
PUSH	R4	2	&H0E
PUSH	R5	2	&H10
PUSH	R6	2	&H12
PUSH	R7	2	&H14
PUSH	R8	2	&H16
PUSH	R9	2	&H18
PUSH	R10	2	&H1A
PUSH	R11	2	&H1C
PUSH	R16	2	&H1E
PUSH	R17	2	&H20
PUSH	R18	2	&H22
PUSH	R19	2	&H24
PUSH	R20	2	&H26
PUSH	R21	2	&H28
PUSH	R22	2	&H2A
PUSH	R23	2	&H2C
PUSH	R24	2	&H2E
PUSH	R25	2	&H30
PUSH	R26	2	&H32
PUSH	R27	2	&H34
PUSH	R28	2	&H36
PUSH	R29	2	&H38
PUSH	R30	2	&H3A
PUSH	R31	2	&H3C
IN	R24,0x3F	1	&H3E
PUSH	R24	2	&H3F
LDI	R24,0x3D	1	&H41
OUT	0x32,R24	1	&H42
IN R24,0x16		1	&H3D
...			

The interrupt occurs after 195 cycles of Timer0. The first activity in the ISR is carried out 66 cycles (&H42) later. Such deviations may be unacceptable in some cases.

Caution: In case of low prescaler values, take into account the time needed for register saving.

Listing 15 shows a simple way (marked in bold) of taking the additional cycles into consideration.

```
Const Reload = 195            ' Reload value for Period of 50 ms
Dim Counter As Byte

Config Timer0 = Timer , Prescale = 1

On Timer0 Timer0_isr          ' Jump to Timer1 ISR

Config Portb = Output

Enable Timer0                 ' Enable the timer interrupt
Enable Interrupts             ' Enable Global Interrupt

Do
     Nop
Loop

End

Timer0_isr:
      ' Reload Timer0 for Period of 50 ms
         Counter = Tcnt0              ' Read Timer0
         Tcnt0 = Counter - Reload     ' Reload Timer0
         Portb.0 = Not Pinb.0         ' Toggle Portb.0
Return
```

Listing 15 Modified clock generation by Timer0 (TIMER0_1.BAS)

Before reloading Timer0, its content is read and the reload value can be corrected before reloading Timer0. The assembler list shows the changes following this program modification.

Instruction		Cycles	TCNT0 (Prescaler=1)
PUSH	R0	2	&H07
PUSH	R1	2	&H09
PUSH	R2	2	&H0B
PUSH	R3	2	&H0D
PUSH	R4	2	&H0F
PUSH	R5	2	&H11
PUSH	R6	2	&H13
PUSH	R7	2	&H15
PUSH	R8	2	&H17
PUSH	R9	2	&H19
PUSH	R10	2	&H1B
PUSH	R11	2	&H1D
PUSH	R16	2	&H1F
PUSH	R17	2	&H21
PUSH	R18	2	&H23
PUSH	R19	2	&H25
PUSH	R20	2	&H27
PUSH	R21	2	&H29
PUSH	R22	2	&H2B
PUSH	R23	2	&H2D
PUSH	R24	2	&H2F
PUSH	R25	2	&H31
PUSH	R26	2	&H33
PUSH	R27	2	&H35
PUSH	R28	2	&H37
PUSH	R29	2	&H39
PUSH	R30	2	&H3B
PUSH	R31	2	&H3D
IN	R24,0x3F	1	&H3F
PUSH	R24	2	&H40
LDI	R26,0x60	1	&H42
LDI	R27,0x00	1	&H43
IN	R24,0x32	1	&H44
ST	X,R24	2	&H45
LDI	R26,0x60	1	&H47
LDI	R27,0x00	1	&H48
LD	R16,X	2	&H49
LDI	R20,0xC3	1	&H4B
SUB	R16,R20	1	&H4C
OUT	0x32,R16	1	&H4D
IN	R24,0x16	1	&H81
...			

The timer period should be 195 cycles of Timer0 again. After 77 cycles (= &H4D) the calculated value of 129 (= &H81) is reloaded. With a prescaler of 1, Timer0 will overflow after 256 - 129 + 77 = 204 cycles.

The remaining difference to the expected value of 195 results from the difference between reading and writing TCNT0 (&H4D-&H44). It is nine cycles here and can be considered when necessary.

4.2.1.2 Counter

In the counter mode the timers/counters of the AVR microcontrollers are able to count (external) events. For Timer0, Pin0 of PortB serves as counter input T0. The leading or falling edge of the input signal can trigger the counter. Register TCNT0 contains the number of counted pulses.

A simple example demonstrates the counter mode of Timer0. What is to be counted are pulse packages of 10 pulses each. The number of received packages will be saved in a variable. Listing 16 shows the resulting source.

```
Const Ticks = 10              ' Number of pulses in a package
Dim Count As Byte             ' Counter

Config Timer0 = Counter , Edge = Falling
Load Timer0 , Ticks           'Overflow Interrupt after 10 cycles

On Timer0 Timer0_isr          ' Jump to Timer0 ISR

Config Portb = Output         ' PortB Output
Reset Ddrb.0                  ' Portb0 Input for pulse counter

Enable Timer0                 ' Enable the timer interrupt
Enable Interrupts             ' Enable Global Interrupt

Do
   Portb = Count * 2          ' Shift one bit left for display
Loop

End

Timer0_isr:
    Load Timer0 , Ticks
    Incr Count
Return
```

Listing 16 Pulse Counter with Timer0 (COUNTER0.BAS)

At the beginning of this program example, Timer0 is configured as counting falling edges of the input signal. Register TCNT0 is loaded so that an overflow interrupt occurs after 10 counted pulses.

The ISR manages reloading and increments variable Count.

In the endless loop, the variable counter is displayed by the LEDs connected to PortB. As Pin0 serves as counter input T0, it is not available for display. The result needs to be shifted one bit to the left, and the rest of PortB is used for display only.

Compared with Timer0, Timer1 offers a lot more features in counter mode as well.

Timer1 can count pulses from Pin1 of PortB (T1). If a capture pulse is detected at Pin4 of PortD (ICP), then the register will be moved to the Capture Register. Pulse edges and noise cancellation can be set by means of instruction Config Timer1 = Counter ....

Listing 17 shows a simple program example intended to run in the simulator. For printing reasons the first line is broken. The whole instruction Config Timer1 = ... must be keyed in in one line.

```
Config Timer1 = Counter , Edge = Falling , Noice Cancel = 1 ,
Capture Edge = Rising
                                   ' Count Input is T1 (PB1)
                                   ' Capture Input is ICP (PD4)

Config Portb = Output              ' Portb0 Output
Config Pinb.1 = Input              ' PinB.1 Input for pulse counter
Portb = 255

Do
    Portb = Icr11 * 4              ' Shift two bits left for display
Loop

End
```

Listing 17   Timer/Counter1 Input Capture (CAPTURE1.BAS)

Since Pin1 of PortB serves as counter input T1, it is not available for display. We have to shift The result must be shifted two bits to the left, and the rest of PortB is used for display only.

It is very important to know the exact result of a configuration like Config Timer1 = Counter , Edge = Falling , Noise Cancel = 1 , Capture Edge = Rising. Take the simulator and inspect the phase of initialization in the single-step mode. Figure 46 shows the content of the Timer1 Register after initialization.

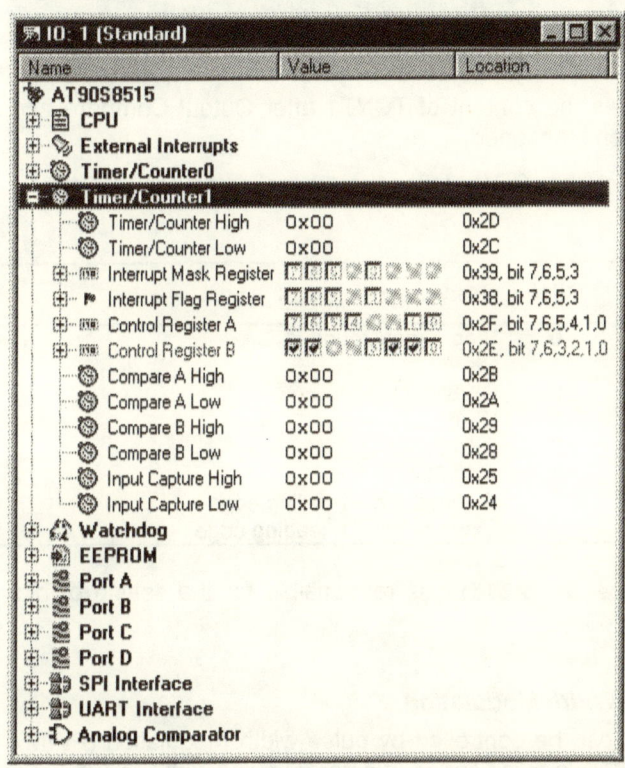

Figure 46   Initial State of TCCR1B in CAPTURE1.BAS

Instruction `Config Timer1 = Counter , Edge = Falling , Noice Cancel = 1 , Capture Edge = Rising` sets the following bits in register TCCR1B:

ICNC1	ICES1		CTC1	CSI2	CSI1	CSI0	
1	1		0	1	1	0	**TCCR1B**

The Input Capture Noise Canceller samples pin ICP four times. Depending on the chosen capture edge, all four samples must be Hi or Lo. Bit ICNC1 must be set (`Noise Cancel = 1`) to activate the noise canceller. If bit ICNC1 is reset (`Noise Cancel = 0`), the noise canceller is deactivated and a single edge will trigger.

The edge for triggering at pin ICP is defined by bit ICES1. If IECS1 is set, the leading edge will trigger (`Capture Edge = Rising`). If

91

ICES1 is reset, the falling edge will trigger (Capture Edge = Falling).

Bit CTC1 defines the content of TCNT1 after Output Compare and has already been mentioned.

The Clock Select1 bits CS1x define the prescaling source of Timer1.

CS12	CS11	CS10	Description
0	0	0	Stop Timer/Counter1
0	0	1	CK
0	1	0	CK/8
0	1	1	CK/64
1	0	0	CK/256
1	0	1	CK/1024
1	1	0	External Pin T1, falling edge
1	1	1	External Pin T1, leading edge

Parameter Edge = Falling is responsible for the setting of bits CS1x in Figure 46.

4.2.1.3 Pulse Width Modulation

A pulse series can be controlled by pulse width modulation (PWM). Figure 47 shows two pulse series of different pulse width or duty.

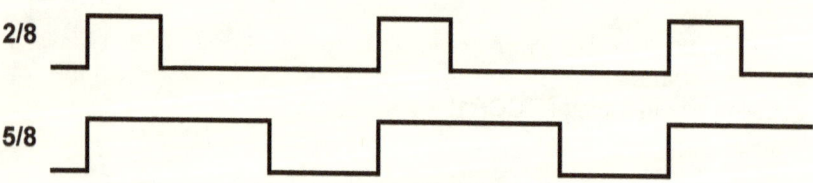

Figure 47 Pulse series of different duties

The upper pulse series has a duty of 2/8 which means that, in a period of eight cycles, two cycles are Hi and the rest of the period is Lo. The lower pulse series has a duty of 5/8.

If such a pulse series is used for driving an LED, the brightness of this LED can be controlled by way of the duty.

Listing 18 shows an example with Timer0 as pulse width modulator. The whole timer period is divided into a Hi and a Lo phase. The ISR has two paths that will be passed through alternatively.

```
Const True = 1
Const False = 0

Dim Hi As Byte
Dim Lo As Byte
Dim A As Byte
Dim Phase As bit
Dim Pattern As Byte               ' bit pattern for display

Pwm Alias Portb.0                 ' Modulated Pin

' Rate 244 Hz at 4 MHz Clock
Config Timer0 = Timer , Prescale = 64

On Timer0 Timer0_isr              ' Jump to Timer0 ISR

Config Portb = Output

Enable Timer0                     ' Enable the timer interrupt
Enable Interrupts                 ' Enable Global Interrupt

Lo = 128                          ' Initial value for PWM
Phase = True

Do
   A = Pind ' Ask for Key
   Select Case A
       Case &B11111110 : Lo = 0    ' Lo Time short
       Case &B11111101 : Lo = 32
       Case &B11111011 : Lo = 64
       Case &B11110111 : Lo = 96
       Case &B11101111 : Lo = 128
       Case &B11011111 : Lo = 160
       Case &B10111111 : Lo = 192
       Case &B01111111 : Lo = 255  ' Lo Time long
   End Select
   Hi = 255 - Lo
   Incr Pattern                   ' Change bit Pattern
   Waitms 100                     ' Wait 100 ms
Loop

End

Timer0_isr:
   If Phase = True Then
       Portb = &HFF               ' LED off
       Timer0 = Lo                ' Reload Timer0
       Phase = False
   Else

       Portb = Not Pattern        ' LED on
```

```
            Timer0 = Hi              ' Reload Timer0
            Phase = True
        End If
Return
```

Listing 18   Brightness Control for LED by PWM (PWM0.BAS)

The following tasks are included in the endless loop of the program:

1. Query the key connected to PortD
2. Set Lo time according to the pressed key to reload Timer0
3. Calculate the corresponding Hi time for reloading Timer0
4. Manipulate the bit pattern for display on PortB
5. Include a waiting time

It is very easy to test this program with the evaluation board. After the start of the program the Lo time is initialized to 128. The LEDs connected to PortB display the changing bit patterns at a mean brightness.

After pressing one of the keys connected to PortD the CASE construct determines a new Lo time, and the brightness of the LED changes. The blinking rate does not change because the Hi time will always be adapted to the changed Lo time.

For PWM, Timer1 offers some more features. Timer1 should be used if a certain precision is expected. In the PWM mode, Timer1 operates as up/down counter comparing TCNT1 with the Output Compare registers OCR1A and OCR1B permanently. If TCNT1 is equal to one of the registers OCR1A or OCR1B, then the actions described next will start.

Digital-to-analog conversion based on PWM is here exemplified by PWM with Timer1. See the register contents for a better understanding. Use the simulator to inspect the initialization process. Listing 19 is a program example.

```
Pwma Alias Portd.5           ' Modulated Pins
Pwmb Alias Oc1b

Dim Temp1 As Word            ' Used Variables
Dim Temp2 As Word

Config Portb = Output        ' PortB is Output
Portb = 255   ' Switch LEDs off
```

```
Config Timer1 = Pwm , Pwm = 10 , Compare A Pwm = Clear Down ,
Compare B Pwm = Clear Up
Temp1 = &H0000                  ' Configure Timer1 for PWM
Pwm1a = Temp1
Pwm1b = Temp1
Tccr1b = Tccr1b Or &H02         ' Prescaler = 8

Config Pind.0 = Input           ' Configure PortD
Config Pind.5 = Output

Do
    bitwait Pind.0 , Reset      ' Wait for key pressed
    bitwait Pind.0 , Set        ' Wait for key unpressed
    Temp1 = Temp1 + &H10        ' Increment Variable
    Pwm1a = Temp1               ' Set PWM Registers
    Pwm1b = Temp1
    Temp2 = Temp1 / &H10        ' Reset 4 LSB and shift right
    Temp2 = Not Temp2           ' Invert bit pattern
    Portb = Low(temp2)          ' Output bit pattern
Loop

End
```

Listing 19   Digital-to-Analog Conversion by PWM (PWM1.BAS)

Instruction Config Timer1 = Pwm , Pwm = 10 , Compare A Pwm = Clear Down , Compare B Pwm = Clear Up manages the setup of register TCCR1A completely. Figure 48 shows the contents of register TCCR1A after configuration.

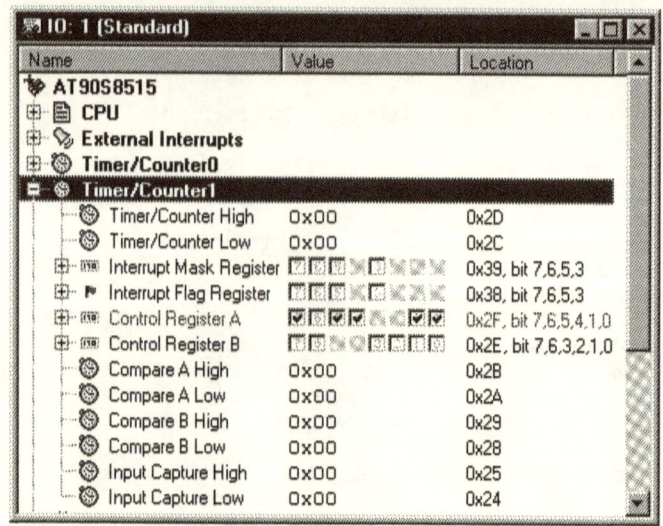

Figure 48   Initialization of TCCR1A

The bits in register TCCR1A are set as follows:

COM1A1	COM1A1	COM1B1	COM1B1			PWM11	PWM10	
1	0	1	1			1	1	TCCR1A

Bits COM1A1 and COM1A0 define the response of output pin OC1A. Bits COM1B1 and COM1B0 define it for OC1B. The x in the next table is to be replaced by A or B.

COM1x1	COM1x0	Description
0	0	not connected
0	1	not connected
1	0	Cleared on compare match, up-counting Set on compare match, down-counting
1	1	Cleared on compare match, down-counting Set on compare match, up-counting

Bits PWM11 and PWM10 define the resolution of PWM to 8 bit, 9 bit or 10 bit.

PWM11	PWM10	Description
0	0	PWM not activated
0	1	8-bit PWM
1	0	9-bit PWM
1	1	10-bit PWM

The resolution defines the counting range of the up/down-counter. The counting range is always from 0 to TOP (see table). The frequency depends on the counting range, too.

Resolution	TOP	PWM Frequency
8 bit	&H00FF	$f_{T/C1}/510$
9 bit	&H01FF	$f_{T/C1}/1022$
10 bit	&H03FF	$f_{T/C1}/2046$

The following equation is used for calculating the PWM frequency:

$$f_{PWM} = \frac{f_{T/C1}}{2^{N+1} - 2}$$

The clock frequency for Timer1 $f_{T/C1}$ is defined by the prescaler and will be set in register TCCR1B.

Figure 49 shows the content of register TCCR1B after running instruction `Tccr1b = Tccr1b Or &H02` at the beginning of this program example.

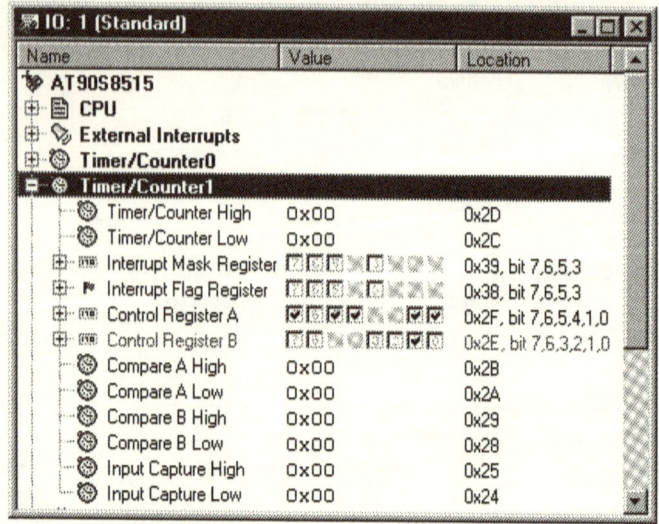

Figure 49 Initialization of TCCR1B

At a clock frequency of 4 MHz a prescaler of eight generates a PWM frequency of about 245 Hz. Connecting a resistor and a capacitor as low pass filter to output OC1A or OC1B is all that is needed to get a simple digital-to-analog converter.

In practice, the low pass can be designed according to this formula:

$$\tau = R \cdot C = \frac{(10..1000)}{f_{PWM}}$$

If time constant τ is too high, the response time will also be high. On the other hand, if time constant τ is too low, the filtering effect will be poor.

Table 3 shows the output voltages measured across pins OC1A/OC1B and ground for program example PWM1.BAS.

To simplify matters, no low pass filter was connected. Due to the integrating measuring principle, the digital multimeter used for measuring had s sufficient filtering capacity.

By pressing the key connected to Pin0 of PortD, the duty of the PWM output can be changed. As there is no debouncing, the change is sometimes greater than expected. The actual duty is displayed by LEDs connected to PortB.

Word	&H000	&H010	&H020	&H030	&H040	&H050	&H060	&H070
PortB	&H00	&H01	&H02	&H03	&H04	&H05	&H06	&H07
OC1A	.001	.078	.156	.234	.311	.389	.467	.544
OC1B	4.93	4.86	4.78	4.70	4.62	4.55	4.47	4.39
Word	&H080	&H090	&H0A0	&H0B0	&H0C0	&H0D0	&H0E0	&H0F0
PortB	&H08	&H09	&H0A	&H0B	&H0C	&H0D	&H0E	&H0F
OC1A	.622	.700	.777	.855	.932	1.01	1.08	1.15
OC1B	4.32	4.24	4.16	4.08	4.01	3.93	3.85	3.77
Word	&H100	&H110	&H120	&H130	&H140	&H150	&H160	&H170
PortB	&H10	&H11	&H12	&H13	&H14	&H15	&H16	&H17
OC1A	1.23	1.31	1.39	1.46	1.54	1.62	1.69	1.77
OC1B	3.70	3.62	3.54	3.47	3.39	3.31	3.23	3.16
Word	&H180	&H190	&H1A0	&H1B0	&H1C0	&H1D0	&H1E0	&H1F0
PortB	&H18	&H19	&H1A	&H1B	&H1C	&H1D	&H1E	&H1F
OC1A	1.85	1.93	2.00	2.08	2.16	2.24	2.31	2.39
OC1B	3.08	3.00	2.93	2.85	2.77	2.69	2.61	2.54
Word	&H200	&H210	&H220	&H230	&H240	&H250	&H260	&H270
PortB	&H20	&H21	&H22	&H23	&H24	&H25	&H26	&H27
OC1A	2.47	2.54	2.62	2.70	2.78	2.85	2.93	3.01
OC1B	2.46	2.39	2.31	2.23	2.15	2.08	2.00	1.92
Word	&H280	&H290	&H2A0	&H2B0	&H2C0	&H2D0	&H2E0	&H2F0
PortB	&H28	&H29	&H2A	&H2B	&H2C	&H2D	&H2E	&H2F
OC1A	3.08	3.16	3.24	3.32	3.39	3.47	3.55	3.63
OC1B	1.85	1.77	1.69	1.61	1.54	1.46	1.38	1.30
Word	&H300	&H310	&H320	&H330	&H340	&H350	&H360	&H370
PortB	&H30	&H31	&H32	&H33	&H34	&H35	&H36	&H37
OC1A	3.70	3.78	3.86	3.93	4.01	4.09	4.17	4.24
OC1B	1.23	1.15	1.07	1.007	.929	.851	.774	.696
Word	&H380	&H390	&H3A0	&H3B0	&H3C0	&H3D0	&H3E0	&H3F0
PortB	&H38	&H39	&H3A	&H3B	&H3C	&H3D	&H3E	&H3F
OC1A	4.32	4.40	4.47	4.55	4.63	4.71	4.78	4.86
OC1B	.618	.541	.463	.385	.308	.230	.152	.075

Table 3   Digital-to-Analog Conversion by PWM

Figure 50 shows the values of Table 3 in a clearly arranged graphic.

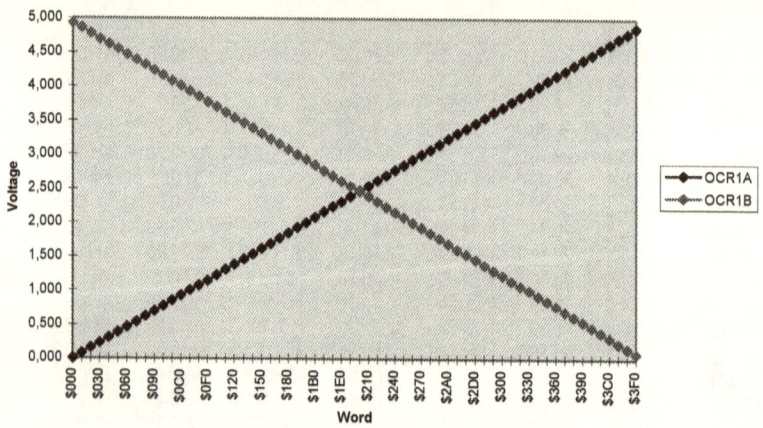

Figure 50   Digital-to-Analog Conversion by PWM

4.2.1.4   Pulse Length Capture

The timers can also be used for capturing the length of a pulse. Figure 51 shows a pulse series with two different Lo phases, t_{p1} and t_{p2}.

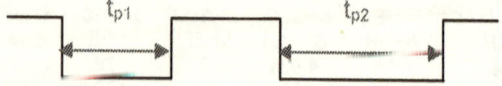

Figure 51   Pulse series

In the simplest case, the timer is started with a falling edge and stopped upon detection of the rising edge. The result in the timer register reflects the measured time. Listing 20 shows program example PULSIN.BAS which uses Timer0 for the time measurement.

```
Declare Function Lopulse() As Byte

Dim Value As Byte

Inputpin Alias Pind.0

Config Portb = Output            ' PortB Output
Portb = &HFF                     ' all LEDs off

Config Timer0 = Timer , Prescale = 1
```

```
On Timer0 Overflow_isr Nosave

Enable Timer0
Enable Interrupts

Do
  Value = Lopulse()
  Portb = Value
Loop

End

Function Lopulse() As Byte
  While Inputpin <> 0 : Wend          ' wait for Hi-Lo on inputpin
  Tcnt0 = 0                            ' reset Timer0
  Start Timer0
  While Inputpin = 0 : Wend
  Stop Timer0                          ' stop Timer0 after 26 cycles minimum
  Lopulse = Tcnt0
End Sub

' overflow isr stops timer0 and set tcnt0 to zero
Overflow_isr:
  !push R24
  Stop Timer0
  Tcnt0 = 0
  !pop R24
  Return
```

Listing 20   Capturing a pulse length (PULSIN.BAS)

The key for capturing the pulse length is function `Lopulse()`. After calling `Lopulse()` the program waits for a falling edge on `Inputpin`. `Inputpin` is an alias for Pin0 of PortD (defined in the third line).

After the detection of a falling edge register TCNT0 is reset and Timer0 starts. Timer0 internally counts clock signals (prescaler = 1) until it is stopped by a rising edge detected on `Inputpin`.

If the pulse is longer than the Timer0 period, a Timer0 Overflow Interrupt occurs. The ISR stops Timer0 and returns to 0.

At a clock frequency of 4 MHz the resolution is (theoretically) 0.25 μs. The run time from detecting the falling edge to detecting the rising edge is 26 cycles, or minimum 6.5 μs. Therefore the capture range is between 6.5 μs and 64 μs.

Using the assembler for edge detection will reduce the runtime. Listing 21 shows the required changes. The changes are marked in bold.

```
Declare Function Lopulse() As Byte

Dim Value As Byte

Const Inputpin = $10 , 0          ' Inputpin Alias Pind.0

Config Portb = Output             ' PortB Output
Portb = &HFF                      ' all LEDs off

Config Timer0 = Timer , Prescale = 1

On Timer0 Overflow_isr Nosave

Enable Timer0
Enable Interrupts

Do
  Value = Lopulse()
  Portb = Value
Loop

End

Function Lopulse() As Byte
  $asm
Hilo:
  Sbic Inputpin                   ' wait for Hi-Lo on inputpin
  Rjmp Hilo
  $end Asm
  Tcnt0 = 0   ' reset Timer0
  Start Timer0
  $asm
Lohi:
  Sbis Inputpin                   ' wait for Hi-Lo on inputpin
  Rjmp Lohi
  $end Asm
  Stop Timer0    ' stop Timer0 after 10 cycles minimum
  Lopulse = Tcnt0
End Sub
```

```
' overflow isr stops timer0 and set tcnt0 to zero
Overflow_isr:
  !push R24
  Stop Timer0
  Tcnt0 = 0
  !pop R24
  Return
```

Listing 21 Capturing a pulse length (PULSIN1.BAS)

The changed function needs ten cycles to detect a rising edge. The minimum pulse length that can be captured is now 2.5 µs.

Capturing a pulse length without an internal timer is demonstrated in Listing 22.

```
Declare Function Lopulse() As Word

Dim Value As Word
Dim Time As Word

Inputpin Alias Pind.0

Config Portb = Output
Portb = &HFF

Do
  Value = Lopulse()
  Portb = Low(value)
Loop

End

Function Lopulse() As Word
  While Inputpin <> 0 : Wend
  Time = 0                       ' reset Time
  While Inputpin = 0
    Incr Time
  Wend
  Lopulse = Time
End Sub
```

Listing 22 Capturing a pulse length (PULSIN2.BAS)

The value of variable Time is proportional to the pulse length. For an exact time specification the cycles of the second while-wend loop are responsible.

At a clock frequency of 4 MHz `Inputpin` is queried every 7.7 μs. The longer sampling time is due to the data formats used for the calculations (word for value and time). An overflow check was not made here.

4.2.2 8051

Most 8051 derivatives have at least two 16-bit timers. These timers are fairly complex circuits. Registers TMOD, TCON and IE control the functionality of these timers.

The timer counts the internal clock divided by 12. The timer period can be calculated according to the following equation:

$$T = 2^N \cdot \frac{12}{f_{OSC}}$$

The timer can operate in four modes:

- Mode 0 : 13-bit timer (8-bit timer with 5-bit prescaler)
- Mode 1 : 16-bit timer.
- Mode 2 : 8-bit timer with auto-reload
- Mode 3 : 8-bit timer (see datasheet for details)

Typically working at a clock frequency of 12 MHz, the timer clock is 1 MHz. For Mode 1 and Mode 2 the following data are obtained:

Timer at 12 MHz clock frequency

Mode	2	1
Maximum Timer Period	256 μs	65.536 ms
Resolution	1 μs	1 μs

Figure 52 shows a block diagram of Timer0 / Timer1. For timer configuration BASCOM-8051 has special instructions that set the respective special function registers (SFR).

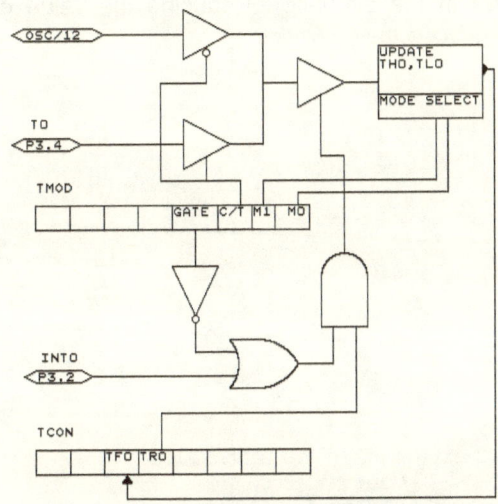

Figure 52 Block diagram of 8051 timer

To configure Timer0, the following bits have to be set/reset in the TMOD and TCON SFRs. At the beginning it is good practice to verify this setup in the simulator.

```
Config Timer0 = Counter      1 -> C/T
Config Timer0 = Timer        0 -> C/T
Config Gate = External       1 -> Gate
Config Gate = Internal       0 -> Gate
Config Mode = 0-3            00-11 -> M1, M0
Start Timer0                 1 -> TR0
Stop Timer0                  0 -> TR0
```

In Listing 23, Timer0 is initialized as timer in mode 2. The timer operates as a reloadable 8-bit timer with a period of 250 µs. When 250 µs are exceeded, the timer overflows and interrupts the program.

Interrupt handler `timer0_isr` increments a counter variable. After 4000 timer interrupts, P3.5 is toggled and the same procedure starts again.

If an INT0 interrupt occurs, the reload value for Timer0 is manipulated, and the blinking rate changes.

In the main loop of the program, P1.7 is toggled every 100 ms to demonstrate some activity of the main program.

This program can be run in the simulator. Reducing the value of Ms_delay has a favorable effect in the simulation.

```
' Timer0 for 8051

Dim Ms_cntr As Integer
Dim Ms_delay As Integer
Dim Rl_value As Byte            ' Reload value for Timer0

' Timer0 is a reloadable 8-bit Timer
Config Timer0 = Timer , Mode = 2

On Timer0 Timer0_isr
On Int0 Int0_isr

Main:
   Rl_value = 250'Timer0 Overflow after 250us at 12 MHz
   Ms_cntr = 0                  'Init Ms_cntr
   Ms_delay = 4000              'Delay of 4000 x 250us = 1000 ms
   Gosub Init_io
   Gosub Init_timer0
   Enable Interrupts            'Global Interrupt Enable
Do
   P1.7 = Not P1.7              'Do Anything Forever
   Waitms 100
Loop

End

'-----------------------------------------------------------
' Interrupt Handler
'-----------------------------------------------------------
Timer0_isr:  'Handler for Timer0 Overflow
   Incr Ms_cntr
   If Ms_cntr = Ms_delay Then
      Ms_cntr = 0
      P3.5 = Not P3.5           'Toggle P3.5
   End If
Return

Int0_isr:
   Rl_value = Rl_value / 2      'Change Rl_value Value
   Stop Timer0
   Load Timer0 , Rl_value       'SetUp Timer0 with changed Rl_value
   Start Timer0
   P3.7 = Not P3.7              'Toggle P3.7
Return

'-----------------------------------------------------------
```

```
' Subroutines
'-----------------------------------------------------------

Init_io:
   P3.7 = 1
   P3.5 = 1
   Set Tcon.0                    'Falling edge triggers INT0
   Enable Int0                   'Enables INT0
Return

Init_timer0:
   Load Timer0 , R1_value        'Store R1_value in Timer0
   Enable Timer0                 'Enable Timer0 Overflow Interrupt
   Start Timer0                  'Start Timer0
Return
```

Listing 23   Timer example (TIMER.BAS)

4.3 LED Control

LEDs or displays based on LEDs are widely used for simple display functions. Their advantage is the excellent visibility. In most cases, however, this advantage must be paid for with a high power consumption.

4.3.1 Single LED

LEDs can be directly driven from the pins of AVR microcontrollers. Due to the electrical specifications it is advantageous to connect the LEDs as is shown in Figure 53.

The following equation is used to calculate the series resistances:

$$R = \frac{V_{CC} - V_{LED} - V_{OL}}{I_{LED}}$$

According to the datasheet of the AT90S8515, the output voltage V_{OL} is 0.6 V maximum at a current of 20 mA.

If an LED is intended to be driven at a current of 10 mA, the series resistance can be determined using the parameters V_{CC} = 5V, V_{LED} = 1,5 V and V_{OL} = 0.3 V (R = 320 Ω).

However, if the resistors shown in Figure 53 are used the current flowing through the LEDs will be lower.

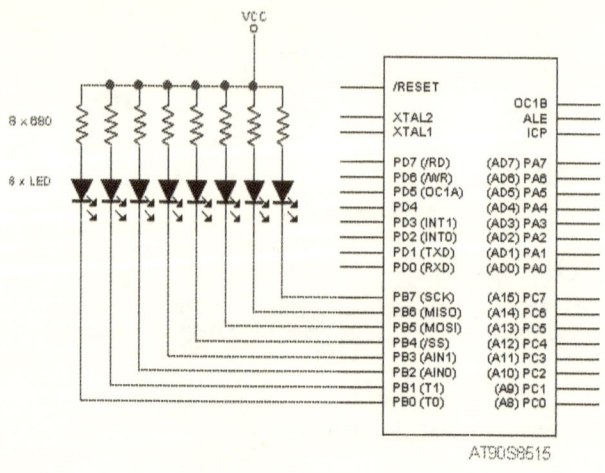

Figure 53 Connecting LEDs to PortB

4.3.2 Seven-Segment Displays

Seven-segment displays can display the figures of our numbering system and a couple of special characters.

There are many types of seven-segment displays from different manufacturers. Basically, this type of display consists of a number of LEDs with connected anodes or cathodes.

In our application example, the type SA03-11 display made by Kingbright is used. Figure 54 depicts such a display.

Figure 54 Seven-Segment Display SA03-11

Caution: To connect an LED display to the port of any microcontroller, adhere to the connecting diagram of the display used.

Figure 55 shows the segment assignment and pin configuration of an SA03-11 display.

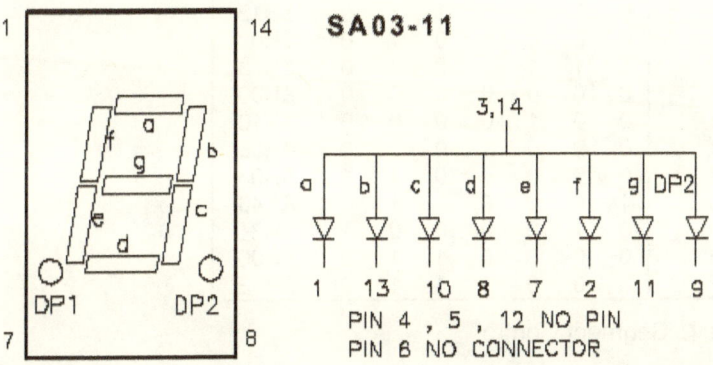

Figure 55 Segment Assignment and Pin Configuration of SA03-11

To display alphanumeric data with such a seven-segment display, it is necessary to define the control scheme.

As can be seen in Figure 55, the anodes of the individual LEDs are interconnected. If a cathode resistor is connected to a microcontroller pin, the LED can be switched on and off. Lo at the controlling pin switches the LED on, and Hi switches it off.

Table 4 shows the segment control for characters 0 to 9 and A to F as is required for displaying hexadecimal numbers.

Character	Segments						Output
	g	f	e	d	c	b	a	
0	1	0	0	0	0	0	0	&H40
1	1	1	1	1	0	0	1	&H79
2	0	1	0	0	1	0	0	&H24
3	0	1	1	0	0	0	0	&H30
4	0	0	1	1	0	0	1	&H19
5	0	0	1	0	0	1	0	&H12
6	0	0	0	0	0	1	0	&H02
7	1	1	1	1	0	0	0	&H78
8	0	0	0	0	0	0	0	&H00
9	0	0	1	0	0	0	0	&H10
A	0	0	0	1	0	0	0	&H08
B	0	0	0	0	0	1	1	&H03
C	1	0	0	0	1	1	0	&H46
D	0	1	0	0	0	0	1	&H21
E	0	0	0	0	1	1	0	&H06
F	0	0	0	1	1	1	0	&H0E

Table 4 Segment Control

Chapter 0 lists the complete character set of a seven-segment display. This table permits a lot more characters to be defined. No more than seven bits of the control byte are required for the complete character set. The MSB can control the decimal point.

Listing 24 shows a program example that displays the characters 0 to F on a seven-segment display continuously.

```
Config Porta = Output          ' PortA Output
Porta = 255                    ' all segments off

Dim I As Byte

Dim X(16) As Byte              ' Array for controlling bit patterns

Restore Value_table

For I = 1 To 16                ' Read data in array
    Read X(i)
Next

Do
    For I = 1 To 16
        Porta = X(i)           ' Display character
        Waitms 250             ' Wait .5 seconds
        Waitms 250
    Next
```

```
Loop

End

Value_table:
Data &H40 , &H79 , &H24 , &H30 , &H19 , &H12 , &H02 , &H78
Data &H00 , &H10 , &H08 , &H03 , &H46 , &H21 , &H06 , &H0E
```

Listing 24 Control of Seven-Segment-Display by AVR
(7SEGMENT.BAS)

The controlling bit patterns (see Table 4) are stored in the ROM in a table named `Value_table`.

Upon program start, the ROM table is copied to array X which makes access to the array (indexed variable) quite simple.

Characters 0 to F are displayed in an endless loop. The two instructions `waitms 250` generate a waiting time of half a second. Two wait instructions are needed because the argument has byte format and is limited to 255!

With the exception of port I/O, program 7SEGMENT.BAS has no AVR specific instructions. To port this program to 8051, all that is required is to adapt the I/O related instructions. Listing 25 shows the modified program for 8051 microcontrollers. The modifications are marked in bold characters.

```
' Seven-segment control by AT89C2051
$sim                            ' comment for normal operation

P1 = 255

Dim I As Byte
Dim X(16) As Byte

Restore Value_table

For I = 1 To 16
  Read X(i)
Next

Do
  For I = 1 To 16
    P1 = Not X(i)               ' inverted for simulation only
    Waitms 250
    Waitms 250
  Next
Loop
```

```
End

Value_table:
Data &H40 , &H79 , &H24 , &H30 , &H19 , &H12 , &H02 , &H78
Data &H00 , &H10 , &H08 , &H03 , &H46 , &H21 , &H06 , &H0E
```

Listing 25 Control of Seven-Segment-Display by 8051
(7SEGMENT.BAS)

In Listing 25 the output instruction was enhanced by the operator not. The reason is the BASCOM-8051 simulator which was used for program testing. Pressing the LCD button causes a display window to appear that contains a seven-segment display, too. Figure 56 shows the open window. As the segments of this display are switched on at Hi, the polarity had to be changed.

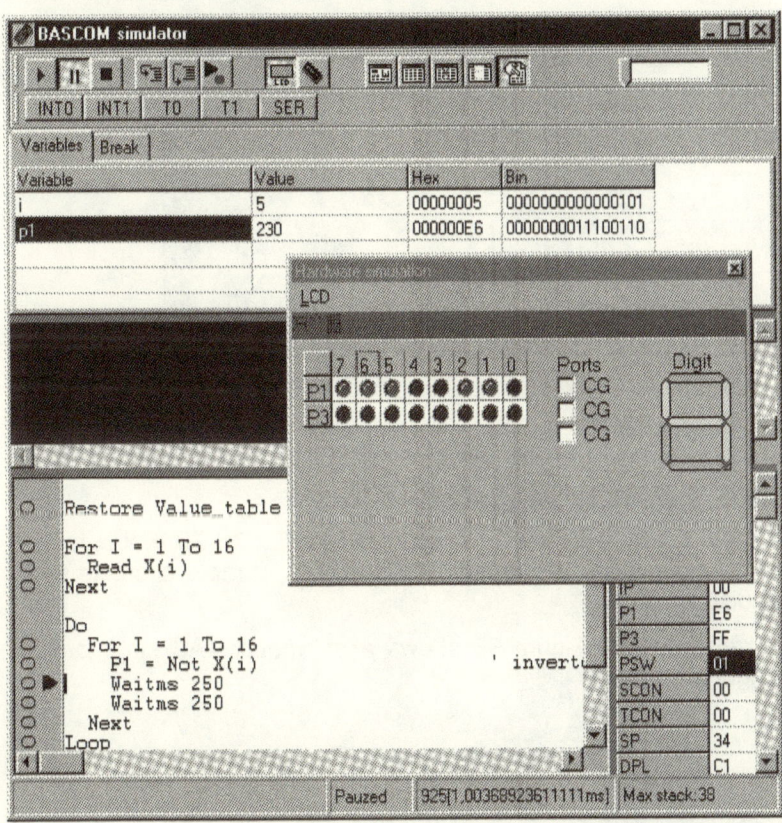

Figure 56 Seven-Segment Display in BASCOM-8051 Simulator

BASCOM-8051 offers the flexibility to assign the segments to any available pin. Right-click the seven-segment display to edit the properties of this display. Figure 57 shows how to edit the digit properties.

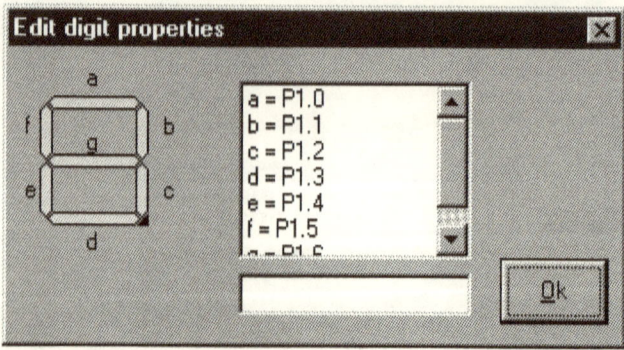

Figure 57   Pin Assignment

4.3.3 Dot-Matrix Displays

In most cases, a dot-matrix display uses a 5 x 7 LED matrix for display purposes. As is common with LCDs, a lot more characters can be displayed.

As an example, Figure 58 shows a dot-matrix TA07-11 made by Kingbright.

Figure 58   Dot-Matrix Display TA07-11

To control such a dot-matrix display, the assignment of these 35 LEDs to the pins of the display must be known. Figure 59 shows the internal circuit diagram of the TA07-11.

Column connections C1 to C5 link up the anodes of all LEDs in a certain column. Row connections R1 to R7 do the same for all LEDs in a certain row.

To switch a LED in the first column and third row, for example, line C1 must be connected to V_{CC} and line R3 via a series resistor to GND.

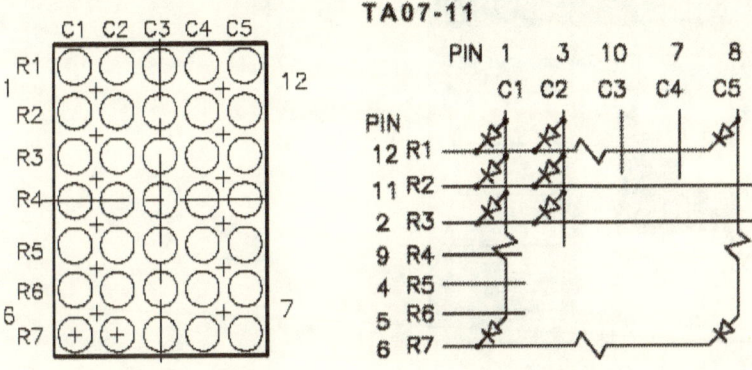

Figure 59 Internal Circuit of Dot –Matrix Display TA07-11

As shown in Figure 59, five column lines and seven row lines are needed to control all LEDs of a 5x7 dot-matrix display. Without extra hardware, each further display needs five additional column lines.

If a dot-matrix display is to be used as a character display, define the characters to be displayed first. Figure 60 shows a graphic character as an example. Let us define these characters next.

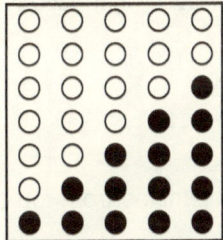

Figure 60 Character to be defined

The LCD Designer, a tool included in BASCOM, can be used not only for LCDs but for this purpose, too. Figure 61 shows the character to be defined with the LCD Designer tool.

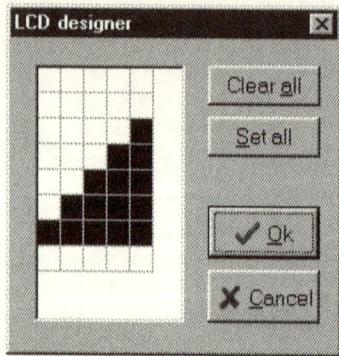

Figure 61 Design of a character

The LCD Designer generates the following instruction for this special character:

```
Deflcdchar ?, 224, 224, 225, 227, 231, 239, 255, 224
' replace ? with number (0-7)
```

Of this instruction only the generated bit patterns are of interest here. These bit patterns are saved in the memory with a DATA instruction as follows:

```
Dotmatrix:
Data 224 , 224 , 225 , 227 , 231 , 239 , 255 , 224
```

These eight bytes describe the bit pattern of the pixel lines from top to bottom. Only five bits of each byte are significant.

The dot-matrix display is driven column after column. Therefore we need bit patterns for columns, not for rows as generated by the LCD Designer. The required conversion can be performed by the micro-controller during initialization.

Figure 62 shows the circuit for driving the dot-matrix display. For more clarity, the circuitry for PortA and PortC is presented only.

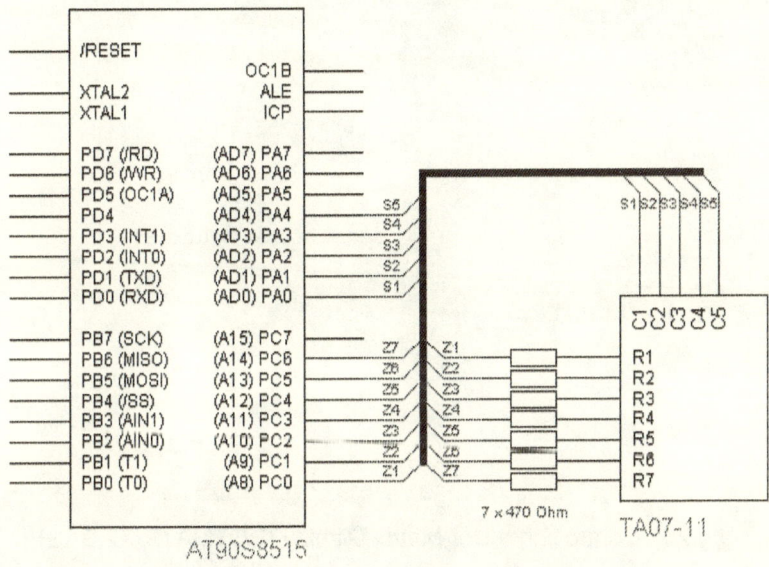

Figure 62 Control Circuit for Dot-Matrix Display TA07-11

Listing 26 is a program example for the display of a character generated by the LCD Designer as described.

```
' Control of Dot-Matrix Display by AVR

Dim A(5) As Byte
Dim I As Byte , J As Byte
Dim X As Byte , Y As Byte

Config Porta = Output            ' all rows Hi
Porta = &HFF

Config Portc = Output            ' all columns Lo
Portc = 0

Restore Dotmatrix

For I = 0 To 7                   ' convert rows to columns
  Read Y                         ' read from table
  Shift Y , Left , 3             ' shift 3 MSB
  X = 0
  For J = 1 To 5
    X = A(j)
    If Y > &H7F Then             ' test for MSB
      Set X.i
```

```
      Else
        Reset X.i
      End If
      A(j) = X
      Shift Y , Left
   Next
Next

Do
  For I = 0 To 4
    X = 0
    Set X.i                        ' set accessed column Hi
    Portc = X
    J = I + 1
    Porta = Not A(j)
  Next
Loop

End

Dotmatrix:
Data 224 , 224 , 225 , 227 , 231 , 239 , 255 , 224
```

Listing 26   Controlling a Dot-Matrix Display (DOTMATRIX2.BAS)

PortA serves as driver for the row lines. PortC drives the column lines of the dot-matrix display. After initialization all LEDs of the display are switched off.

The conversion of the bit pattern from pixel rows to pixel columns starts after resetting the data pointer to the first data byte of the bit patterns. Because the three most significant bits of each pixel row are not needed, they are cancelled by instruction Shift Y , Left , 3. As shown in the next table, the remaining five bits in each pixel row are inspected column by column. So the pixel positions in variables A(1) to A(5) will be set or reset bit by bit starting at the LSB.

224	1	1	1	0	0	0	0	0
224	1	1	1	0	0	0	0	0
225	1	1	1	0	0	0	0	1
227	1	1	1	0	0	0	1	1
231	1	1	1	0	0	1	1	1
239	1	1	1	0	1	1	1	1
255	1	1	1	1	1	1	1	1
224	1	1	1	0	0	0	0	0
				A(1)	A(2)	A(3)	A(4)	A(5)

At a clock frequency of 4 MHz the whole conversion process takes about 2.2 ms. This time will not be noticed during initialization.

An endless loop drives the dot-matrix display by outputting the converted bit pattern column after column.

For enhancing the display to several devices either more column driver lines (five for each device) or extra hardware for multiplexing are required.

4.4 LCD Control

LCDs are receiving advanced features for the display of information. The number of low-priced LCDs offered on the market is immense.

Fortunately, the HD44780 LCD controller by Hitachi or compatible devices are used in most cases for alphanumeric displays.

Basically, it is distinguished between two kinds of device control. In the direct mode the pins of the microcontroller drive the lines of the connected LCD directly. In the other case, some LCDs are equipped with a standardized RS232 or I^2C interface. The number of required interface lines decreases. For small microcontrollers, it is often the latter aspect that is of importance.

4.4.1 Direct Control

The LCD controller type HD44780 provides the connected microcontroller with an 8-bit bus and a number of control lines. The pins of such an LCD module have the following meanings:

Pin	Designation	Level	Function
1	V_{SS}	GND	GND
2	V_{DD}	+5 V	Supply voltage
3	Vo	0 ... +5 V	Contrast control
4	RS	H/L	L: Instruction register H: Data register
5	R/W	H/L	L: Read access H: Write access
6	E	H/L	Enable
7 - 14	DB0 - DB7	H/L	Data lines

There are two ways to connect a microcontroller in the direct mode. See Figure 14 for configuring the connection mode.

If the microcontroller circuit works with an external memory or memory-mapped I/O, then a data bus exists and the LCD can be connected in the bus mode. The SetUp of the STK200 evaluation board has already been shown in Figure 14. Figure 63 depicts the connection of an LCD module with LCD controller HD44780 to the data bus and the control lines of an AT90S8515 microcontroller.

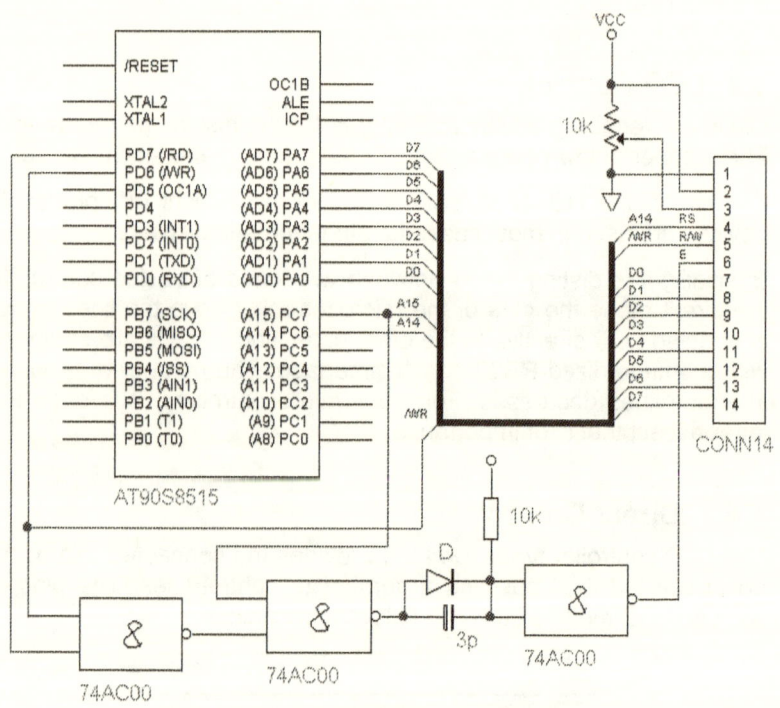

Figure 63   LCD Connected to AT90S8515 in Bus Mode

The LCD controller type HD44780 has two internal 8-bit registers that can be accessed from the connected microcontroller.

The instruction register (IR) saves the received commands (RS = 0). The data register (DR) saves data (RS = 1) which are sent to the Data Display RAM (DD RAM) or Character Generator RAM (CG RAM). Address line A14 distinguishes between instructions and data.

Together with the Read/Write signals, address line A15 controls the Enable line of the LCD module. A falling edge at the Enable input (E) of the LCD controllers latches the data (D7-D0).

If there is no external bus the LCD can be connected in the pin mode which means the SetUp must assign the pins of the LCD to the corresponding pins of the microcontroller.

The below table shows possible assignments:

LCD	Pin	Port
DB7	14	PORTB.7
DB6	13	PORTB.6
DB5	12	PORTB.5
DB4	11	PORTB.4
E	6	PORTB.3
RS	4	PORTB.2
RW	5	GND
V_{ss}	1	GND
V_{dd}	2	+5 Volt
Vo	3	0-5 Volt

In this configuration PORTB.1 and PORTB.0 (and the other Ports not used here) are available for other purposes.

After correct initialization in the LCD SetUp the LCD can be controlled using comfortable LCD instructions. Listing 27 shows a simple LCD control program for a first test.

```
' LCD Control by AVR and 8051
$sim                          ' for simulation only otherwise comment

Dim A As Byte

M1:
  A = Waitkey()

  If A = 27 Then Goto M2
  Cls
  Upperline
  Lcd A
  Lowerline
  Lcd Hex(a)                  ' uncomment for AVR
'  Lcdhex A                   ' uncomment for 8051
  Print Chr(a)
  Goto M1
M2:
  End
```

Listing 27   LCD Control (LCD.BAS)

The program waits for a character to be sent. If the character sent is ESC the program will end. Otherwise, the display (16 characters, 2 lines) shows the received character on both lines in different formats.

If one prefers to go inside BASCOM, then the internal routines can be used, too. The following is an example for BASCOM-AVR.

```
$ASM
Ldi _temp1, 5                'load register R24 with value
Rcall _Lcd_control           'it is a control value
                             'to control the display

Ldi _temp1,65                'load register with new value (letter A)
Rcall _Write_lcd             'write it to the LCD-display
$END ASM
```

Subroutines _lcd_control and _write_lcd are written in assembler and can be called from BASIC.

4.4.2 LCD with Serial Interface

LCDs with a serial interface offer a simplified connectivity. In the simplest case two wires (TxD & GND) from the microcontroller to the LCD are sufficient.

The comfortable LCD instruction cannot be used for this kind of LCD control. Some knowledge of the LCD controller is required.

As the DD RAM of the HD44780 LCD controllers has 80 bytes, one HD44780 LCD controller can control one LCD with four lines of max. 20 characters each.

Table 5 shows the LCD position and DD RAM address for a 4x16 LCD (LM041L etc.) as an example.

DD RAM	1	2	3	4	5	6	7	8	9	10	11	12	13	14	15	16
1. Zeile	00	01	02	03	04	05	06	07	08	09	0A	0B	0C	0D	0E	0F
2. Zeile	40	41	42	43	44	45	46	47	48	49	4A	4B	4C	4D	4E	4F
3. Zeile	10	11	12	14	13	15	16	17	18	19	1A	1B	1C	1D	1E	1F
4. Zeile	50	51	52	53	54	55	56	57	58	59	5A	5B	5C	5D	5E	5F

Table 5   Display position and DD RAM address for LCD 4x16

As shown in Table 5, not all memory space is used for display in a 4x16 LCD. DD RAM not used for the display is available as external RAM. The access to that external RAM requires a complete RS-232. Table 6 shows an extract from the instruction set of an HD44780LCD controller. Table 7 describes some designations in Table 6.

Instruction	RS	DB7	DB6	DB5	DB4	DB3	DB2	DB1	DB0	Description
Clear Display	0	0	0	0	0	0	0	0	1	Clears display and sets cursor to home position
Cursor At Home	0	0	0	0	0	0	0	1	X	Sets cursor to home position
Set Entry Mode	0	0	0	0	0	0	1	I/D	S	Defines direction of cursor and shift movements
Display On/Off	0	0	0	0	0	1	D	C	B	See explanations of D, C, and B
Cursor/Display Shift	0	0	0	0	1	S/C	R/L	X	X	See explanations of S/C and R/L
Function Set	0	0	0	1	DL	N	F	X	X	See explanations of DL, N, and F
Set CG RAM Addr	0	0	1	ACG						Sets CG RAM address
Set DD RAM Addr	0	1	ADD							Sets DD RAM address
Data Write	1	Data								Writes byte in DD RAM or CG RAM

Table 6 Coding of Instructions of LCD Controller HD44780

Name	Description
I/D	After writing a character to RAM, the DD RAM or CG RAM address will be incremented (I/D = 1) or decremented (I/D = 0).
S	Moving the contents of display to the right (S = 1) or left (S = 0) Cursor position does not change (calculator).
D	Display on (D = 1) or off (D = 0). Data in DD RAM remain unchanged.
C	Cursor on (C = 1) or off (D = 0).
B	Cursor blinking (B = 1) or not blinking (B = 0).
S/C	Moves the contents of display (S/C = 1) or the cursor (S/C = 0) by one position according to R/L.
R/L	Moving to the right (R/L = 1) or left (R/L = 0) without changes in DD RAM.
DL	Data bus 8 bit (DL = 1) or 4 bit (DL = 0).
N	Number of display lines - one (N = 0) - several (N = 1).
F	Font - 5 x 7 dots (F = 0) - 5 x 10 dots (F = 1).
X	Don't care.

Table 7 Explanation of instruction set of HD44780 LCD controllers

After this short description of the basics of LCD controller type HD44780, the next program example can be interpreted.

Listing 28 shows a BASCOM-AVR program controlling an LCD via RS-232. Due to the serial interface all commands for the LCD controller must be sent by print instructions from the microcontroller.

This program example is based on a serial LCD from Scott Edwards Electronics [http://www.seetron.com]. Scott's web site offers a lot of information about all types of LCDs.

```
'--------------------------------------------------
' SW_UART.BAS for AVR
' Controlling a LCD with LCD Serial Backpack from
' SEETRON
' C. Kuehnel
' 1999-11-21
'--------------------------------------------------

Const Instr = 254

Const Clr = 1

Const Lcd_blank = 8
Const Lcd_restore = 12

Const Line1 = &H80
Const Line2 = &HC0
Const Line3 = &H94
Const Line4 = &HD4

Dim I As Byte

Config Portd = Input
Portd = 255

Open "COMC.0:2400,8,N,1,inverted" For Output As #1

Print #1 , Chr(instr) ; Chr(clr);
Print #1 , Chr(instr) ; Chr(line1);
Print #1 , "BASCOM-AVR writes to";

Print #1 , Chr(instr) ; Chr(line2);
Print #1 , "4 line/20 column LCD";

Print #1 , Chr(instr) ; Chr(line3);
Print #1 , "via serial interface";

Print #1 , Chr(instr) ; Chr(line4);
Print #1 , "from www.seetron.com";

Wait 5
```

```
While Pind.0 = 1
    Print #1 , Chr(instr) ; Chr(lcd_blank);
    Waitms 200 : Waitms 200
    Print #1 , Chr(instr) ; Chr(lcd_restore);
    Wait 1 : Waitms 250 : Waitms 250
Wend

Close #1

End
```

Listing 28   Control of a Serial LCD (SW_UART.BAS)

A number of constants are declared in the first part of the program.

PortD is initialized as input because the program looks for Pin0 of PortD to run or end.

Before any display the serial interface must be initialized, too. No complete serial interface is needed for this program example; the transmit line (TxD) will do. Therefore, the UART software is good enough for this purpose.

The asynchronous serial communication will be discussed in chapter 4.7. You may read this chapter to get first information, or accept the initialization of the UART software with `Open "COMC.0:2400,8,N,1,inverted" For Output As #1` and the data output with `Print #1 , ...` first. The data are output from Pin0 of PortC of the microcontroller used at 2400 baud and with inverted polarity.

We have to distinguish between two types of data output:

```
Print #1 , Chr(instr) ; Chr(line1);
Print #1 , "BASCOM-AVR writes to";
```

In the first instruction, data byte `instr` announces a command (RS=0). The command itself is the data byte `line1` (DD RAM = 00) which sets the data pointer to the first position in DD RAM.

The second instruction transfers data to be displayed to the DD RAM, starting at the preselected location.

Before entering the while-wend loop all data are written to the LCD. In the loop, the display is cleared and reactivated (Restore) periodically. Because the arguments for instructions `Wait` and `Waitms` are limited to one byte, several wait instructions need to be added to generate longer wait times.

In BASCOM-8051 it is different. The UART software only supports the GET and PUT statements, and the PRINTBIN and INPUTBIN statements to retrieve and send data. It is not possible to simply send a string with `print "abcdefg"`. COM1 and COM2 are hardware ports that can be used with PRINT etc.

Listing 29 shows the LCD control program ported to BASCOM-8051. Pin0 of Port1 serves as key input. The UART hardware sends the data to the serially connected LCD.

```
'---------------------------------------------------
' Serial_LCD.BAS for 8051
' Controlling a LCD with LCD Serial Backpack from
' SEETRON
' C. Kuehnel
' 2001-01-01
'---------------------------------------------------

Const Command = 254

Const Clr = 1

Const Lcd_blank = 8
Const Lcd_restore = 12

Const Line1 = &H80
Const Line2 = &HC0
Const Line3 = &H94
Const Line4 = &HD4

Dim I As Byte
Dim Key As Bit

Open "COM1:2400,inverted" For Output As #1        ' RS232 inverted!

Print #1 , Chr(command) ; Chr(clr);
Print #1 , Chr(command) ; Chr(line1);
Print #1 , "BASCOM-AVR writes to";

Print #1 , Chr(command) ; Chr(line2);
Print #1 , "4 line/20 column LCD";

Print #1 , Chr(command) ; Chr(line3);
Print #1 , "via serial interface";

Print #1 , Chr(command) ; Chr(line4);
Print #1 , "from www.seetron.com";

Wait 5

Do
    Print #1 , Chr(command) ; Chr(lcd_blank);
```

```
       Waitms 200 : Waitms 200
       Print #1 , Chr(command) ; Chr(lcd_restore);
       Wait 1 : Waitms 250 : Waitms 250
       Key = P1.0
Loop Until Key = 0

Close #1

End
```

Listing 29   Control of a Serial LCD (SERIAL_LCD.BAS)

4.5 Connecting Keys and Keyboards

Keyboards for microcontrollers need not have the same features as keyboards for PCs. Often simple keypads as shown in Figure 64 are sufficient for input purposes.

This kind of keypad is available in two types:

- 1x12 single keys
- 3x4 key matrix

Figure 64   Keypad

If there are enough I/O lines available then the single key version can be used. If not, three I/O lines can be saved when the key matrix is used.

4.5.1 Single Keys

In the single key version (1x12) the keypad shown in Figure 64 has the internal connections shown in Figure 65.

All keys of the keypad are wired up on one side and connected to pin1. The other side of each key is connected to one of the pins 2 to 13.

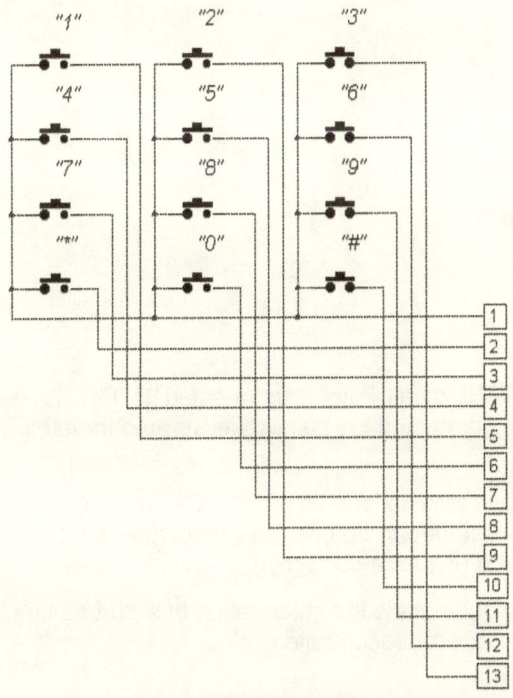

Figure 65   Internal Wiring of Keypad 1x12

Figure 66 shows how to connect such a keypad to a microcontroller. To simplify the diagram the keypad was reduced to four keys.

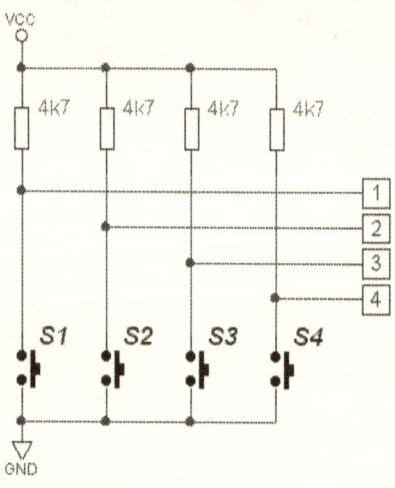

Figure 66   Reduced Keypad

The wired end (COMMON) of all keys is connected to GND. The other side of each key is connected via pull-up resistors to the supply voltage.

Most microcontrollers have internal pull-up resistors at their I/O ports. These internal pull-up resistors can be used to reduce the number of components on your board (initialize correctly!).

The next two lines of the code show the initialization of a port as input with internal pull-ups for AVR microcontrollers.

```
Config Porta = Input      ' Porta is input
Porta = 255               ' with internal pull-up
```

The first line initializes the data direction registers for input, and the second line sets the data register to Hi to enable the pull-up resistors.

Pressing a key generates a falling edge at the respective I/O pin, and the microcontroller can detect this event.

It is important to consider the bouncing of all kinds of mechanical keys. Debouncing is no issue under BASCOM: `debounce` is a very helpful instruction. Listing 30 shows the query of the reduced keypad using the `debounce` instruction.

```
' Query a keypad by AVR

Const Keys = 4                     ' Test for 4 keys only

Config Portb = Output              ' Portb is output

Config Porta = Input               ' Porta is input
Porta = 255                        ' with internal pull-up
Config Portc = Input               ' PortC is input
Portc = 255                        ' with internal pull-up

Dim I As Byte
Dim Key As Byte                    ' Variable contains key number

Portb = 255                        ' Switch LEDs off

Do
   For I = 1 To Keys               ' Query all keys
      Key = I
      Select Case Key
         Case 1 : Debounce Pina.0 , 0 , Display_key , Sub
         Case 2 : Debounce Pina.1 , 0 , Display_key , Sub
         Case 3 : Debounce Pina.2 , 0 , Display_key , Sub
         Case 4 : Debounce Pina.3 , 0 , Display_key , Sub
      End Select
   Next
Loop

End

Display_key:
    Portb = Not Key                ' Display key number by LED
Return
```

Listing 30 Query of a Keypad by AVR (KEY1.BAS)

Listing 31 shows the slightly modified program for an 8051 microcontroller. The differences result from the differing port I/O only.

```
' Query a keypad by 8051

Const Keys = 4                     ' Test for 4 keys only

Dim I As Byte
Dim Key As Byte                    ' Variable contains key number

' Port1 is drives LEDs
P1 = 255                           ' Switch LEDs off

' Port2 is input with internal pull-up
```

```
P2 = 255                        ' needed for input

Do
 For I = 1 To Keys              ' Query all keys
  Key = I
  Select Case Key
   Case 1 : Debounce P2.0 , 0 , Display_key , Sub
   Case 2 : Debounce P2.1 , 0 , Display_key , Sub
   Case 3 : Debounce P2.2 , 0 , Display_key , Sub
   Case 4 : Debounce P2.3 , 0 , Display_key , Sub
  End Select
 Next
Loop

End

Display_key:
    P1 = Not Key                ' Display key number by LED
Return
```

Listing 31 Query of a Keypad by 8051 (KEY1.BAS)

4.5.2 Matrix Keypad

If there is only a limited number of I/O lines available for the keypad, a matrix keypad will be the better solution.

Figure 67 shows the changed internal wiring for the same keypad. The single keys are wired up in the form of columns and rows, with pins 1 to 3 connecting the columns and pins 4 to 7 the rows.

Pressing key "1" connects pin 1 and pin 7, for example.

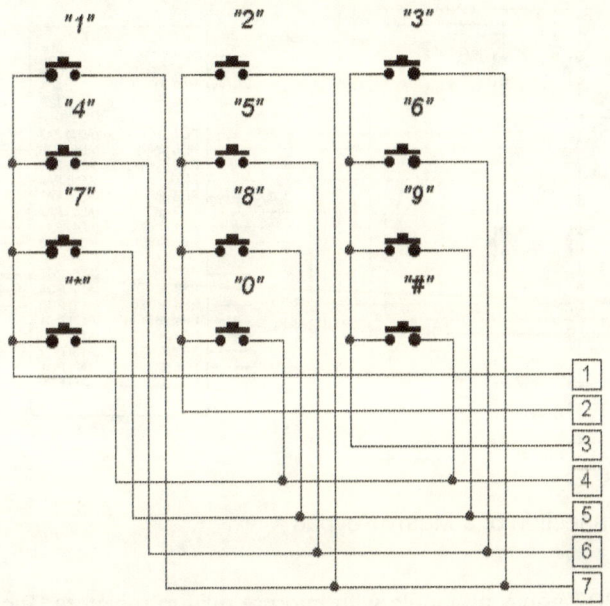

Figure 67 Internal Wiring of a 3x4 Keypad

Figure 68 shows how to connect a matrix keypad to an AVR microcontroller when internal pull-up resistors are used.

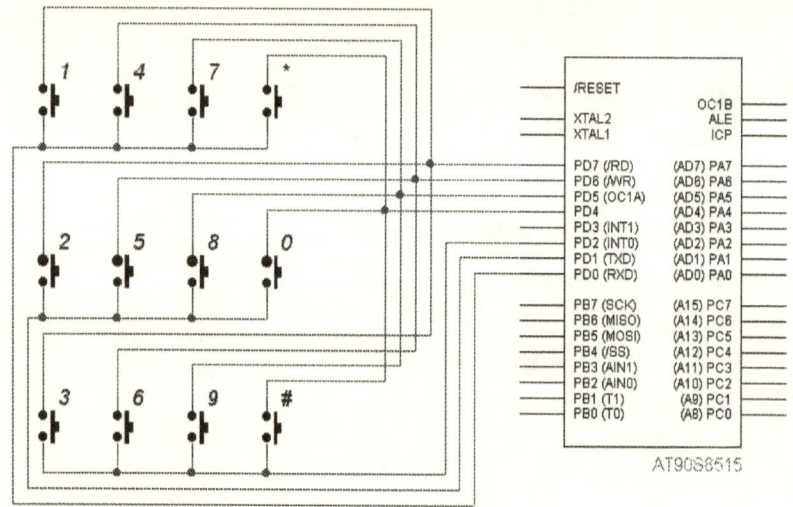

Figure 68   Connection of a Matrix Keypad

Pins PD4 to PD7 serve as inputs with internal pull-up resistors. Pins PD0 to PD2 set the queried column line to Lo.

A query of a matrix keypad divides into several queries of key columns. Listing 32 shows two interlocked loops to query this matrix keypad for an AVR microcontroller.

```
' Query a matrix keypad by AVR

Config Portb = Output

Ddra = &H0F                     ' PD7-PD4 Input; PD3-PD0 Output
Porta = &HFF                    ' with internal pull-up

Dim Column As Byte
Dim Row As Byte
Dim Key As Byte                 ' Variable contains key number

Portb = 255                     ' Switch LEDs off

Do                              ' Query all keys
  For Column = 0 To 2
    If Column = 0 Then Reset Porta.0
    If Column = 1 Then Reset Porta.1
    If Column = 2 Then Reset Porta.2
    'If Column = 3 Then Reset Porta.3
    For Row = 4 To 7
      Select Case Row
```

```
            Case 4 : Debounce Pina.7 , 0 , Calc_key , Sub
            Case 5 : Debounce Pina.6 , 0 , Calc_key , Sub
            Case 6 : Debounce Pina.5 , 0 , Calc_key , Sub
            Case 7 : Debounce Pina.4 , 0 , Calc_key , Sub
        End Select
    Next
    Porta = &HFF
  Next
Loop

End

Calc_key:
  Select Case Row
    Case 4 : Key = Column + 1
    Case 5 : Key = Column + 4
    Case 6 : Key = Column + 7
    Case 7 : Key = Column + 10
  End Select
  Portb = Not Key
Return
```

Listing 32   Matrix Keypad Query (KEY2.BAS)

In the inner loop `For Row = 4 to 7 ... Next` the keys of one column will be queried. The column to be queried is activated in the outer loop `For Column = 0 To 2 ... Next` by resetting the respective I/O pin.

To query a 4x4 matrix (hex keypad, for example) the column query must be changed to `For Column = 0 To 3 ... Next`.

Additionally, BASCOM-AVR has the function `GETKBD()` for querying a 4x4 matrix keypad. See the BASCOM-AVR help for the required details.

Instruction `Config Kbd = Porta` assigns any port of the AVR to the matrix keypad. A keypad query now needs one function call only. The next program lines show how easy it is to encode a keypad input.

```
Config Kbd = Porta
Config Portb = Output

Dim Value As Byte

Value = Getkbd()
Portb = not Value
```

4.5.3 PC-AT Keyboard

PC-AT keyboards are nowadays offered at low prices and can be used in microcontroller applications which don't use the whole functionality. Old PC-XT keyboards have a different functionality and won't be dealt with here.

The PC-AT keyboard sends a scan code when a key is pressed or released. The BIOS of the PC evaluates this scan code.

Pressing key "A", for example, causes the keyboard to send the scan code &H1C (Make Code). If the key is kept pressed, the keyboard will send this scan code again after a defined time. This procedure repeats as long as that key is pressed, or another key is pressed.

After releasing the key, the keyboard sends the scan code &HF0 followed by &H1C (Break Code). The Break Code differs from the Make Code by the leading byte &HF0.

As shown in Figure 69, each key has its own scan code. Whether the Shift key needs to be pressed is determined by the PC BIOS. In the same way the PC BIOS controls the LEDs in the keyboard when one of the keys Num Lock, Caps Lock or Scroll Lock is pressed.

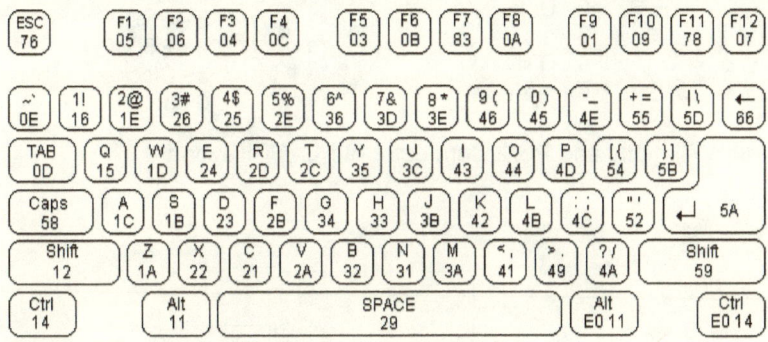

Figure 69 Scan Codes of a PC-AT Keyboard

It is, however, definitely wrong to think that the 101 keys of a PC-AT keyboard generate 101 different scan codes in byte format.

Some keys are so-called Extended Keys. Their scan codes have a leading &HE0. Pressing key Pause generates the following sequence of scan codes: &HE1, &H14, &H77, &HE1, &HF0, &H14, &HF0, &H77!

Since the microcontroller is not supported by a BIOS, the scan codes must be decoded in the application program. BASCOM-AVR supports querying the PC-AT keyboard by function Getatkbd().

Before discussing the software, it is worthwhile to have a closer look at the hardware interface between PC-AT keyboard and microcontroller.

Figure 70 shows the available connectors for a PC-AT keyboard (DIN and PS/2 connector). As can be seen from pinout, the data exchange is synchronous and serial. The data line is bidirectional. The clock is always generated by the PC-AT keyboard.

Figure 71 shows the simple interfacing of a PC-AT keyboard to an AT90S8515. Any pin can be chosen for this kind of interface.

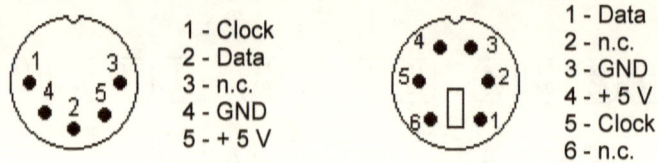

Figure 70 DIN and PS/2 Connector of PC-AT Keyboard

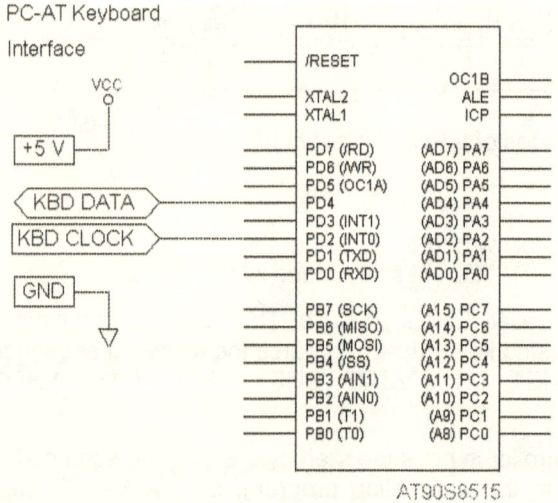

Figure 71 Interface between PC-AT Keyboard and AVR

Function `Getatkbd()` is here used for querying the PC-AT keyboard. Listing 33 shows a simple program example for this purpose.

```
' Query a PC-AT keyboard by AVR

Config Keyboard = Pind.2 , Data = Pind.4 , Keydata = Keydata

Dim B As Byte

Print "hello"

Do
  B = Getatkbd()                    'get a byte and store it into
                                    'byte variable
  'When no real key is pressed the result is 0
  'So test if the result was > 0
  If B > 0 Then
      Print B ; Chr(b)
  End If
Loop
End

'This is the key translation table

Keydata:
'normal keys lower case
Data 0 , 0 , 0 , 0 , 0 , 0 , 0 , 0 , 0 , 0 , 0 , 0 , 0 , 0 , &HBE , 0
Data 0 , 0 , 0 , 0 , 0 , 113 , 49 , 0 , 0 , 0 , 122 , 115 , 97 , 119 , 50 , 0
Data 0 , 99 , 120 , 100 , 101 , 52 , 51 , 0 , 0 , 32 , 118 , 102 , 116 , 114 , 53 , 0
Data 0 , 110 , 98 , 104 , 103 , 121 , 54 , 7 , 8 , 44 , 109 , 106 , 117 , 55 , 56 , 0
Data 0 , 44 , 107 , 105 , 111 , 48 , 57 , 0 , 0 , 46 , 45 , 108 , 48 , 112 , 43 , 0
Data 0 , 0 , 0 , 0 , 0 , 92 , 0 , 0 , 0 , 0 , 13 , 0 , 0 , 92 , 0 , 0
Data 0 , 60 , 0 , 0 , 0 , 0 , 8 , 0 , 0 , 49 , 0 , 52 , 55 , 0 , 0 , 0
Data 48 , 44 , 50 , 53 , 54 , 56 , 0 , 0 , 0 , 43 , 51 , 45 , 42 , 57 , 0 , 0

'shifted keys UPPER case
Data 0 , 0 , 0 , 0 , 0 , 0 , 0 , 0 , 0 , 0 , 0 , 0 , 0 , 0 , 0 , 0
Data 0 , 0 , 0 , 0 , 0 , 81 , 33 , 0 , 0 , 0 , 90 , 83 , 65 , 87 , 34 , 0
Data 0 , 67 , 88 , 68 , 69 , 0 , 35 , 0 , 0 , 32 , 86 , 70 , 84 , 82 , 37 , 0
Data 0 , 78 , 66 , 72 , 71 , 89 , 38 , 0 , 0 , 0 , 76 , 77 , 74 , 85 , 47 , 40 , 0
Data 0 , 59 , 75 , 73 , 79 , 61 , 41 , 0 , 0 , 58 , 95 , 76 , 48 , 80 , 63 , 0
Data 0 , 0 , 0 , 0 , 0 , 96 , 0 , 0 , 0 , 0 , 13 , 94 , 0 , 42 , 0 , 0
Data 0 , 62 , 0 , 0 , 0 , 8 , 0 , 0 , 49 , 0 , 52 , 55 , 0 , 0 , 0
Data 48 , 44 , 50 , 53 , 54 , 56 , 0 , 0 , 0 , 43 , 51 , 45 , 42 , 57 , 0 , 0
```

Listing 33 Query of a PC-AT Keyboard (ATKBD.BAS)

Before the PC-AT keyboard can be queried, the used pins must be assigned and a table containing the scan codes must be prepared.

Pin2 of PortD receives the clock from the PC-AT keyboard while Pin4 serves as data line.

Function Getkbd() queries the PC-AT keyboard for data by analyzing the bit stream received from the keyboard (Figure 72). This bit stream contains the scan codes.

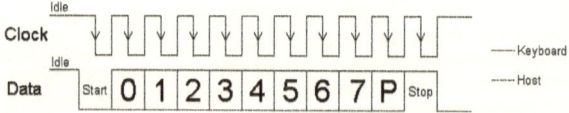

Figure 72 Bit Stream from Keyboard Controller

The connected microcontroller circuitry serves as power supply for the PC-AT keyboard. Note the power consumption of the PC-AT keyboard used. In case it is not known, measure the power supply current of the PC-AT keyboard used to avoid a damaging of the power supply.

4.6 Data Input by IR Remote Control

Most audio or video systems available today have IR remote controls for user interaction. Widespread are remote controls manufactured by SONY or Philips which operate with a standardized transmission protocol (RC5).

The RC5 protocol consists of a 14-bit data word. The data word uses the so-called Manchester coding, a bi-phase code. Figure 73 shows an IR command according to RC5.

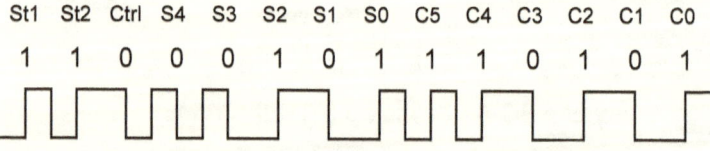

Figure 73 RC5 Coded IR Command

The command begins with two start bits (St1, St2) which are always set. The following bit (Ctrl) toggles for each command. Repeated commands can be detected this way. The control bit is followed by five system bits (S4-S0). The control bits contain the address of the device to be controlled. Usually, TV sets have an address of 0, video recorder an address of 5, etc. Six command bits (C5-C0) close the sequence. There are 64 different commands available for each device.

Table 8 shows an extract from a list of devices and their RC5 addresses.

System	Device	
0	Video	TV1
1		TV2
5		VCR1
6		VCR2
17	Audio	Tuner
20		CD
21		Phono
18		Recorder1

Table 8 RC5 Device Address

In correspondence with these explanations, the command in Figure 73 sends command value &H35 to VCR1.

```
' Query an IR Remote Control by AVR

Const Tv = 0                       ' TV address is 0

Config Rc5 = Pind.2                ' Configures PinD.2 as RC5 Input
Portd.2 = 1                        ' Activates Pull-up

Enable Interrupts                  ' Getrc5 uses timer0 interrupt

Dim Address As Byte , Command As Byte

Do
   Getrc5(address , Command)       ' Query IR remote control

   If Address = Tv Then             ' Check for the TV address
      Print Address ; " " ; Command
   End If
Loop

End
```

Listing 34 Query of an IR Remote Control (RC5.BAS)

As shown in Listing 34, function GETRC5() handles the whole RC5 protocol.

What about the hardware? Siemens offers the IR receiver SFH506-36 for this purpose. It is very simple to connect this receiver to a microcontroller. Figure 74 shows the connection of an IR receiver SFH506 to an AVR. It is important that the internal pull-ups are activated.

In consideration of the different port situation between the 8051 and the AVR, program RC5.BAS can be modified to suit BASCOM-AVR without any problems.

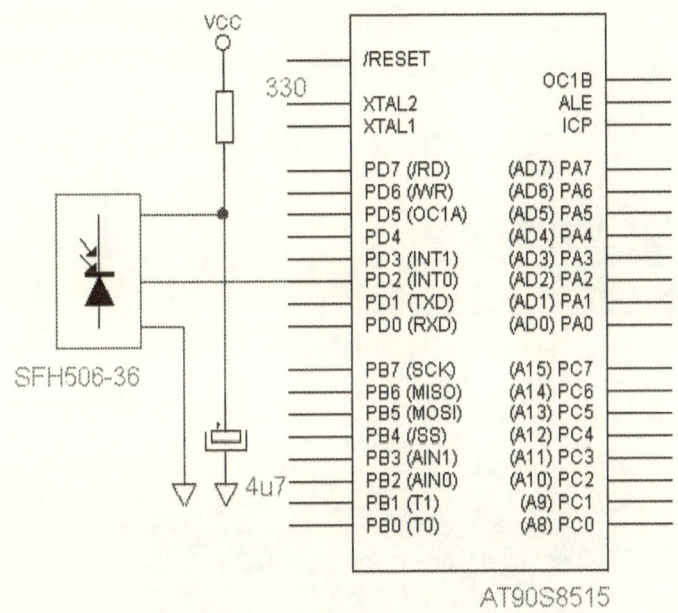

Figure 74   SFH506 connected to AT90S8515

4.7 Asynchronous Serial Communication

For asynchronous serial communication, microcontrollers use an internal UART (Universal Asynchronous Receiver Transmitter), or it must be implemented in the software (emulation).

In BASCOM, the I/O instructions common in BASIC (input and print) are redirected to the serial port. That means instruction Input reads characters from and instruction Print sends characters to the serial port.

A signal converter is required to connect a microcontroller to the COM port of a PC. Well-known are MAX232 or compatible devices. Figure 75 shows a MAX231 connected to the UART pins of an AT90S8515. The MAX231 is equivalent to MAX232 but needs one capacitor only.

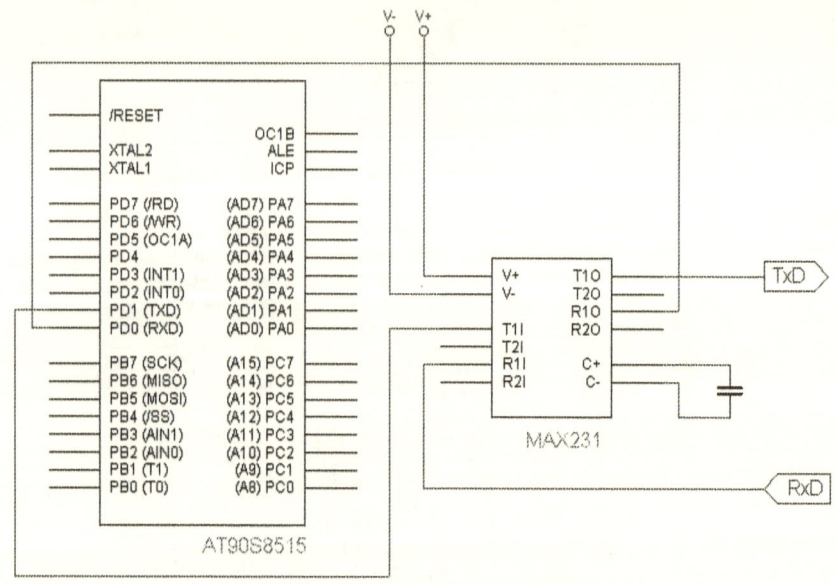

Figure 75 RS-232 Level Conversion by MAX231

For communication, use the BASCOM-internal Terminal Emulator or any other terminal program at the PC end. Program RS232MON, which can be downloaded from authors' web site, can be used for debugging at byte level.

BASIC instructions `Input` and `Print` are also available in BASCOM. Listing 35 shows how to use them.

```
' Serial I/O by AVR and 8051

Dim A As Integer

Do
  Input "Input Number: " , A
  Print " Number was " ; A
Loop

End
```

Listing 35   Serial I/O (SERIAL1.BAS)

Variable A is declared as integer. Figure 76 shows the conversion of that number to the range of integer numbers -32768 ... 32767.

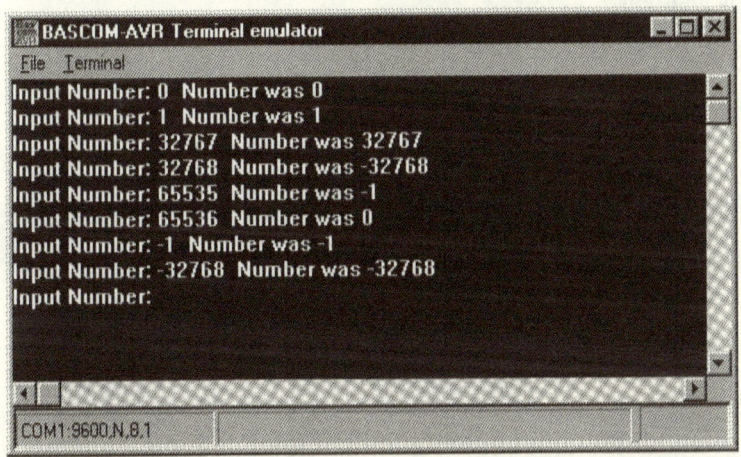

Figure 76 Input of an Integer

Very important for character input without finishing Carriage Return (CR) is instruction `Inputbin`. Listing 36 shows a simple program example.

```
' Serial I/O by AVR and 8051

Dim A As Byte
Dim B As Word                ' B is a reserved word for BASCOM-8051

Do
  Inputbin A , B
  Printbin A , B             ' use Printbin A ; B for BASCOM-8051
  Print
Loop

End
```

Listing 36 Character I/O (SERIAL2.BAS)

Variable A is declared as byte and variable B as word. Instruction `Inputbin` waits for three characters (bytes) without a CR.

In BASCOM-8051, B is a reserved word. Therefore, the name of this variable must be changed for BASCOM-8051. Moreover, the syntax of instruction `Printbin` is different. See the remark in Listing 36.

After receiving three bytes, instruction Printbin sends these three bytes back. The output by Printbin is completed by a CR/LF output by instruction Print.

Related to this kind of input are the functions Inkey() and Waitkey(). Waitkey() waits until a character is received, while Inkey() reads one character from the input buffer. Both functions store that character in a variable. If the input buffer is empty, Inkey() hands over the value of 0. Listing 37 shows a program example and Figure 77 the respective outputs in the terminal window.

```
' Serial I/O by AVR and 8051

Dim A As Byte

Do
  A = Waitkey()                    ' waits for one character
  Print Chr(a) ; " is ASCII " ; A
Loop Until A = 27

Do
  A = Inkey()                      ' reads one character
  Print Chr(a) ; " is ASCII " ; A
  Waitms 100
Loop Until A = 27

End
```

Listing 37   Input by Waitkey() and Inkey() (SERIAL3.BAS)

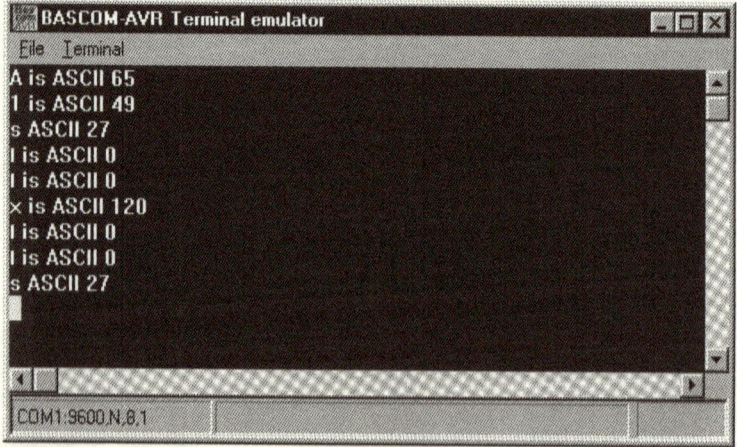

Figure 77   Input by Waitkey() and Inkey()

146

The first loop containing function `Waitkey()` will be reached after program start.

Upon pressing key A on the PC keyboard, the Terminal Emulator sends character A to the connected microcontroller. `Waitkey()` receives this character as expected. The procedure is repeated when key 1 is pressed. The print instruction echoes the character and its ASCII code to the Terminal Emulator each time this is done.

Pressing key ESC quits this first loop. The program progresses to the second loop containing the function `Inkey()`. In this loop, function `Inkey()` will return 0 as long as one character is received by the serial port and written in the input buffer. In Figure 77, character x was received before leaving the loop by pressing ESC again. A wait time of 100 ms slows down the passing of the loop.

Beside serial communication with the internal UART of the microcontroller used, there is the possibility of using an UART emulation. Instructions OPEN and CLOSE serve to configure the communication channels.

Instruction OPEN initializes the communication channel by assigning a pin for input or output and selecting a baud rate.

These examples show the opening of a communication channel for serial output:

```
' Open for AVR
Open "COMA.0:9600,8,N,1,INVERTED" For Output As #1

'Open for 8051
Open "COM3.0:9600,8,N,1,INVERTED" For Output As #1
```

For AVR, Pin0 of PortA is opened as serial output (transmitter) with a baud rate of 9600 Baud, one stop bit and inverted polarity of the RS-232 signal.

For 8051, Pin0 of Port3 was initialized with the same parameters.

Each communication channel that is open at any time must be closed by instruction CLOSE before the end of the program.

In the next program example, two microcontrollers will be serially connected.

Figure 78 shows the circuit diagram of those two linked microcontrollers. Used here are an AT90S8515 and a BASIC Stamp II (BS2). For information on the BS2 see the Appendix.

Instead of a BS2, any microcontroller with a serial port, or a PC running a terminal program, can be used.

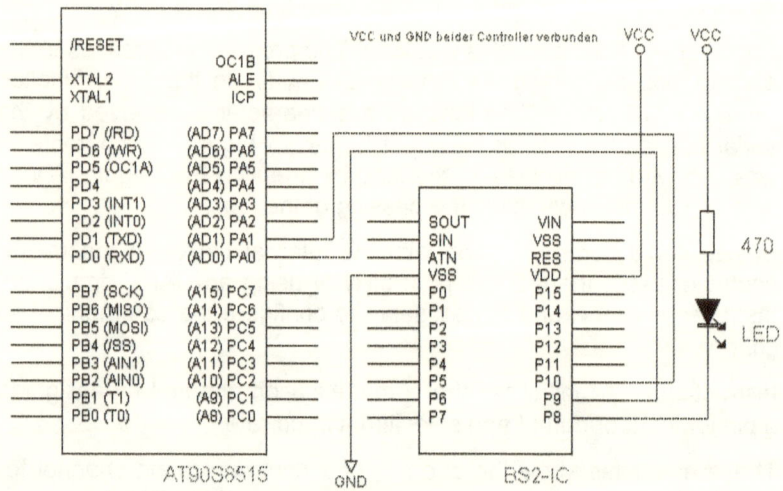

Figure 78 Coupling of AT90S8515 and BASIC Stamp II

The AVR microcontroller uses its internal UART for communication with a terminal as usual. Pins PD0 and PD1 are connected to a MAX232 for level conversion. This part of the circuit is not shown in Figure 78.

In this program example, the second serial interface connected to the BS2 is of importance. Pins PA0 and PA1 form the communication channel for the UART software.

At the BS2 end, pins P9 and P10 serve as serial interface. Pin P8 drives an LED for signalization.

Listing 38 shows the program for the AVR microcontroller and Listing 39 that for the BS2.

If it is intended to replace the AVR by an 8051 derivative, remember to make the required changes in the source code:

- B is a reserved word in BASCOM-8051; replace it by another term for the name of the variable
- Modify the port designation

```
' SW UART by AVR

Dim A As Byte
Dim B As Byte

Open "COMA.0:2400,8,N,1,inverted" For Output As #1
Open "COMA.1:2400,8,N,1,inverted" For Input As #2

Do
  Print "Input one character: ";
  A = Waitkey()
  Print Chr(a)
  Print " Sent character = " ; Chr(a)
  Printbin #1 , A
  Inputbin #2 , B
  Print " Received character = " ; Chr(b)
  Print
Loop Until A = 27

Close #1
Close #2

End
```

Listing 38 AVR Software UART (SERIAL4.BAS)

After the configuration of the transmitter and receiver for serial communication, the program will pass the loop as long as the ESC key is pressed.

Function Waitkey() waits for a character from the terminal and sends it - after some terminal outputs - to the connected BS2 (`Printbin #1 , A`). Thereafter, the program waits for a character from BS2 (`Inputbin #1 , A`) and sends it to the terminal. Figure 79 shows the dialog in the Terminal Emulator.

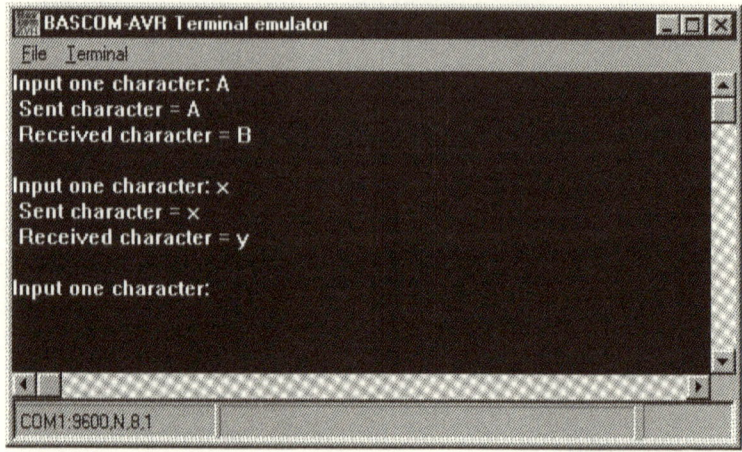

Figure 79 Dialog in Terminal Emulator

As expected, the character input to the terminal is sent to BS2. BS2 increments the received character and sends it back.

Listing 39 shows the small BS2 program. This short program is easy to understand even without any BS2-specific knowledge.

```
LED con 8                        ' Pin8 controls the LED
RxD con 9                        ' Pin9 is receiver
TxD con 10                       ' Pin10 is transmitter

baud con 396+$4000               ' 2400 Baud inverted polarity

char var byte

start:serin RxD, baud, [char]    ' receive one character
      low LED
      pause 500                  ' flash LED
      high LED
      char = char +1             ' increment received character
      serout TxD, baud , [char]  ' send one character
      goto start
end
```

Listing 39 BS2 Transceiver (SERIAL.BS2)

In an endless loop instruction serin RxD, baud, [char] waits for a character to be received at Pin9. The LED connected to Pin8 flashes for 500 ms and the received character is incremented before it is sent back by instruction serout TxD, baud, [char] via Pin10.

By using software UARTs, the microcontrollers can be equipped with several serial ports.

4.8 1-WIRE Interface

Dallas Semiconductors developed the 1-wire interface to reduce the required wiring for networking peripheral components. For data exchange between different components only one wire is needed.

Figure 80 shows the bus master and one slave in a 1-wire network communicating via one line with a pull-up resistor of 4.7 kΩ. The communication occurs in a time-slice procedure hidden to the BASCOM user. The BASCOM instructions guarantee the required timing.

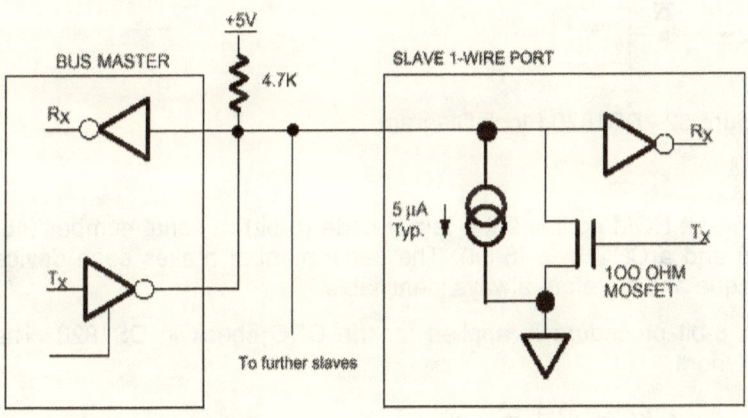

Figure 80   1-Wire Bus System

The following program examples elucidate applications of the 1–Wire Digital Thermometer DS1820 without any exception (Figure 81). This device offers a number of interesting features; its selection is not purely coincidental.

Figure 82 shows the functionality of the DS1820 in a block diagram.

Figure 81   DS1820

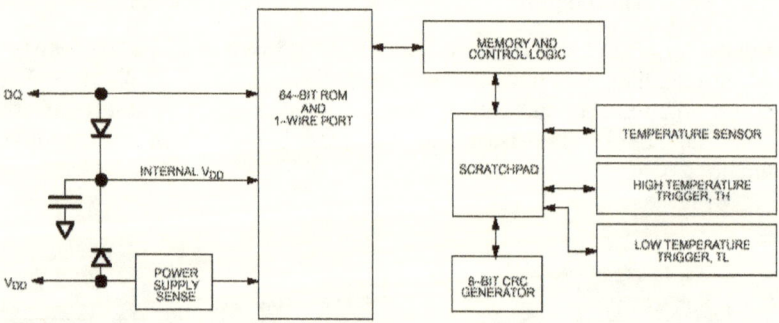

Figure 82   DS1820 Block Diagram

A 64-bit ROM contains the family code (8-bit), a serial number (48-bit) and a CRC byte (8-bit). The serial number makes each device unique and therefore always identifiable.

An 8-bit procedure is applied for the CRC check in DS1820. The polynom

$$CRC = x^8 + x^5 + x^4 + 1$$

is the base of this CRC check.

The DS1820 measures the ambient temperature in a range from -55 °C to +125 °C in increments of 0,5 °C. The temperature value has an internal resolution of 9 bit; see the next table.

Temperature	Binary value	Hex Value
+ 125°C	&B0000000011111010	&H00FA
+ 25°C	&B0000000000110010	&H0032
+ 1/2°C	&B0000000000000001	&H0001
+ 0°C	&B0000000000000000	&H0000
− 1/2°C	&B1111111111111111	&HFFFF
− 25°C	&B1111111111001110	&HFFCE
− 55°C	&B1111111110010010	&HFF92

To fully exploit the accuracy of the DS1820, the temperature is exactly calculated as follows:

1. Read the temperature value and clear the LSB (TEMP_READ)
2. Read the internal counter (COUNT_REMAIN)
3. Read Counts/°C (COUNTS_PER_C)
4. Calculate the temperature value according to the next formula:

$$TEMPERATURE = TEMP\_READ - 0.25 + \frac{COUNT\_PER\_C - COUNT\_REMAIN}{COUNT\_PER\_C}$$

An alarm flag is set when the measured temperature exceeds the threshold TH (or TL).

As long as the alarm flag is set, the DS1820 responds to the Alarm Search command. In a network with several DS1820 devices a simple query for a temperature alarm is enough, i.e. it is not necessary to query each DS1820 separately.

A so-called scratchpad RAM supports the data exchange. The temperature thresholds are saved in a non-volatile EEPROM. The next table shows a memory map of the internal RAM in DS1820.

RAM	Byte	EEPROM
Temperature LSB	0	
Temperature MSB	1	
TH/User Byte 1	2	TH/User Byte 1
TH/User Byte 2	3	TH/User Byte 2
Reserved	4	
Reserved	5	
Count Remain	6	
Count per °C	7	
CRC	8	

The following commands support communication in a network with several DS1820 devices. For detailed information on the DS1820 see the data sheet.

Refer to the program example for further details.

ROM COMMANDS

Read ROM	Reads the complete ROM (possible for DS1820 only)
Match ROM	Addresses a DS1820 by means of the 64-bit ROM content
Skip ROM	Skip addressing (possible for DS1820 only)
Search ROM	Search for DS1820 in a network
Alarm Search	Search for DS1820 in a network reporting an alarm

MEMORY COMMANDS

Convert Temperature	Starts measuring temperature
Read Scratchpad	Reads scratchpad memory
Write Scratchpad	Stores the temperature threshold in the scratchpad memory
Copy Scratchpad	Copies the temperature threshold into the EEPROM
Recall EE	Copies the temperature threshold back to the scratchpad memory
Read Power Supply	Queries the supply voltage

In the next example, a DS1820 is connected to Pin0 of PortA of an AVR microcontroller. Note the pull-up resistor: without it a 1-wire interface will not work!

Listing 40 shows how the connected DS1820 is identified by reading the ROM. To modify this program for the 8051, all that needs to be done is change the ports.

```
' DS1820 Control by AVR

Const Read_rom = &H33          ' DS1820 Commands
Const Skip_rom = &HCC
Const Convertt = &H44
Const Read_ram = &HBE
Const Write_ram = &H4E
Const Copy_ram = &H48
Const Recall_ee = &HB8
Const Read_power = &HB4

Const Slow = 255
Const Fast = 50

Dim I As Byte                  ' Index
Dim Rate As Byte               ' Blink rate

Dim Crc As Byte                ' DS1820 CRC

' Serial Number of DS1820 Device
Dim Serial_number(6) As Byte

Dim Family_code As Byte        ' DS1820 Family Code = &H10

Config Portb = Output          ' Portb is output
Portb = 255

Config 1wire = Porta.0         ' Config PortA.0 as 1wire pin

1wreset                        ' 1wire Reset
If Err = 1 Then                ' On Error blink fast
   Rate = Fast
   Goto Blink
End If

1wwrite Read_rom               ' Read ROM command

Family_code = 1wread()         ' Read 8 Bytes ROM contents
For I = 1 To 6
   Serial_number(i) = 1wread()
Next
Crc = 1wread()
```

155

```
1wreset                                 ' 1wire Reset
If Err = 1 Then                         ' On Error blink fast
    Rate = Fast
    Goto Blink
End If

' Display Family Code
Portb = Not Family_code : Wait 1

' Display 6-Byte Serial Number
For I = 1 To 6
    Portb = Not Serial_number(i) : Wait 1
Next
Portb = Not Crc : Wait 1                ' Display CRC

Rate = Slow : Goto Blink                ' On End blink slow
End

Blink:                                  ' Portb.0 blinks on error
Do
    Portb.0 = 1
    Waitms Rate
    Portb.0 = 0
    Waitms Rate
Loop
```

Listing 40
Reading Family Code and Serial Number (1WIRE1.BAS)

New in Listing 40 are only the 1-wire instructions. At the beginning of the program any 1-wire commands used are declared as constants.

Subroutine Blink serves as an indicator of the state of operation. Variable `Rate` defines a different blinking rate of the connected LED. A fast blinking LED indicates an error.

Pin0 of PortA is the I/O line for the 1-wire interface.

Instruction `1wreset` resets the 1-wire bus. Variable `Err` indicates whether the 1-wire bus reacts as expected or not. On error the program branches to the blink routine.

The next instruction, `1wwrite Read_rom`, informs the DS1820 that its ROM will be read next. Since only one DS1820 is connected to the 1-wire bus, it can be directly addressed.

For several DS1820 in a network the 64-bit address of the DS1820 to be accessed must first be sent to the network.

The eight instructions `1wread()` read the family code, serial number and CRC from the connected DS1820.

156

A further bus reset closes the whole operation. If this bus reset is done without any error occurring, all network operations were errorless.

The program finishes by displaying the read data on PortB. Slow blinking indicates an errorless end of the program.

The temperature measuring procedure is quite similar to the last program example. Listing 41 shows the program source for AVR microcontrollers.

Regarding the modification required to adjust to the BASCOM-8051, the conditions are the same as those mentioned before.

```
' DS1820 Control by AVR

Const Read_rom = &H33              ' DS1820 Commands
Const Skip_rom = &HCC
Const Convertt = &H44
Const Read_ram = &HBE
Const Write_ram = &H4E
Const Copy_ram = &H48
Const Recall_ee = &HB8
Const Read_power = &HB4

Const Slow = 255
Const Fast = 50

Dim I As Byte                      ' Index
Dim Rate As Byte                   ' Blink rate
Dim Busy As Byte

Dim Scratch(9) As Byte             ' Sceatchpad

Config Portb = Output              ' Portb is output
Portb = 255                        ' LEDs off

Config 1wire = Porta.0             ' Config PortA.0 as 1wire pin

1wreset                            ' 1wire Reset
If Err = 1 Then                    ' On Error blink fast
    Rate = Fast
    Gosub Blink
End If

1wwrite Skip_rom                   ' Read ROM command
1wwrite Convertt                   ' Measure Temperature
Do
    Busy = 1wread()
Loop Until Busy = &HFF             ' Wait for end of conversion
```

```
1wreset                           ' 1wire Reset
If Err = 1 Then                   ' On Error blink fast
    Rate = Fast
    Gosub Blink
End If

1wwrite Skip_rom                  ' Skip ROM command
1wwrite Read_ram                  ' Read Scratch command

For I = 1 To 9                    ' Read 9 Bytes Scratch contents
    Scratch(i) = 1wread()
Next

1wreset                           ' 1wire Reset
If Err = 1 Then                   ' On Error blink fast
    Rate = Fast
    Gosub Blink
End If

' Display Temperature LSB
Portb = Not Scratch(1) : Wait 1
' Display Temperature MSB
Portb = Not Scratch(2) : Wait 1

Rate = Slow : Goto Blink          ' On End blink slow
End

Blink:                            ' Portb.0 blinks on error
Do
    Portb.0 = 1
    Waitms Rate
    Portb.0 = 0
    Waitms Rate
Loop
Return
```

Listing 41 DS1820 Temperature Measurement (1WIRE2.BAS)

Basically, the temperature measurement program is quite similar to the last program. Additionally, however, one array of nine bytes serving as mirror for the scratchpad memory of the DS1820 is declared.

Only one DS1820 is connected to the AVR microcontroller as agreed. Therefore, after resetting the 1-wire bus, addressing can be skipped and the temperature measurement can be started immediately.

The end of conversion is detected by the repeated reading of the DS1820. The next instructions prepare the DS1820 for reading the scratchpad RAM.

After reading the scratchpad RAM its content is saved in the declared array byte by byte. The result of the temperature measurement is written to PortB (LSByte first, MSByte second).

The error indication – a blinking LED – does not differ from that of the last program example.

The DS1820 uses a simple 8-bit CRC check for data security. To secure data transmission, the master has to check the received data as well.

Listing 42 shows a program example operating an 8-bit CRC check. The 256 values, or possible results, are saved in a table at the end of the program. To avoid misunderstanding – what is spoken of here is 16 DATA instructions of 16 data bytes each.

```
' This procedure calculates the cumulative Dallas
' Semiconductor 1-Wire CRC of all bytes passed to it.
' The result accumulates in the global variable CRC.

Const Rate = 100            ' Blink rate

Dim Idx As Byte
Dim I As Byte , J As Byte
Dim Crc As Byte             ' Global CRC
Dim X As Byte               ' Input variable
Dim Z(8) As Byte            ' Data bytes

Declare Sub Calc_crc(byval X As Byte)

Config Portb = Output       ' Portb is output
Portb = 255                 ' LEDs off

Crc = 0                     ' Reset CRC

Z(1) = &H02                 ' Eight data bytes for CRC check
Z(2) = &H1C
Z(3) = &HB8
Z(4) = &H01
Z(5) = &H00
Z(6) = &H00
Z(7) = &H00
Z(8) = &HA2
```

```
For J = 1 To 8
    X = Z(j)                    ' Initialize input variable
    Call Calc_crc(x)            ' Calculate CRC
    Portb = Not Crc             ' Display CRC
    Wait 1                      ' Wait a little bit, comment later
Next

Wait 1                          ' Wait for end , comment later

Do                              ' On End blink slowly
    Portb.0 = 1
    Waitms Rate
    Portb.0 = 0
    Waitms Rate
Loop

End

sub Calc_crc(byval X As Byte)
    Restore Crc_table
    Idx = Crc Xor X
    For I = 0 To Idx
        Read Crc
    Next
End Sub

Crc_table:
Data 0 , 94 , 188 , 226 , 97 , 63 , 221 , 131 , 194 , 156 , 126 , 32 , 163 , 253 , 31 , 65,
Data 157 , 195 , 33 , 127 , 252 , 162 , 64 , 30 , 95 , 1 , 227 , 189 , 62 , 96 , 130 , 220,
Data 35 , 125 , 159 , 193 , 66 , 28 , 254 , 160 , 225 , 191 , 93 , 3 , 128 , 222 , 60 , 98,
Data 190 , 224 , 2 , 92 , 223 , 129 , 99 , 61 , 124 , 34 , 192 , 158 , 29 , 67 , 161 , 255,
Data 70 , 24 , 250 , 164 , 39 , 121 , 155 , 197 , 132 , 218 , 56 , 102 , 229 , 187 , 89 , 7,
Data 219 , 133 , 103 , 57 , 186 , 228 , 6 , 88 , 25 , 71 , 165 , 251 , 120 , 38 , 196 , 154,
Data 101 , 59 , 217 , 135 , 4 , 90 , 184 , 230 , 167 , 249 , 27 , 69 , 198 , 152 , 122 , 36,
Data 248 , 166 , 68 , 26 , 153 , 199 , 37 , 123 , 58 , 100 , 134 , 216 , 91 , 5 , 231 , 185,
Data 140 , 210 , 48 , 110 , 237 , 179 , 81 , 15 , 78 , 16 , 242 , 172 , 47 , 113 , 147 , 205,
Data 17 , 79 , 173 , 243 , 112 , 46 , 204 , 146 , 211 , 141 , 111 , 49 , 178 , 236 , 14 , 80,
Data 175 , 241 , 19 , 77 , 206 , 144 , 114 , 44 , 109 , 51 , 209 , 143 , 12 , 82 , 176 , 238,
Data 50 , 108 , 142 , 208 , 83 , 13 , 239 , 177 , 240 , 174 , 76 , 18 , 145 , 207 , 45 , 115,
Data 202 , 148 , 118 , 40 , 171 , 245 , 23 , 73 , 8 , 86 , 180 , 234 , 105 , 55 , 213 , 139,
Data 87 , 9 , 235 , 181 , 54 , 104 , 138 , 212 , 149 , 203 , 41 , 119 , 244 , 170 , 72 , 22,
Data 233 , 183 , 85 , 11 , 136 , 214 , 52 , 106 , 43 , 117 , 151 , 201 , 74 , 20 , 246 , 168,
Data 116 , 42 , 200 , 150 , 21 , 75 , 169 , 247 , 182 , 232 , 10 , 84 , 215 , 137 , 107 , 53
```

Listing 42   Calculation of 8-bit CRC (1WIRE3.BAS)

For test purposes, array Z(8) contains eight bytes simulating eight bytes received from the scratchpad memory. These eight bytes are checked and the result is displayed at PortB byte by byte.

When this program is simulated, the following sequence is displayed at PortB: &H43, &H50, &HE1, &H23, &H0B, &HEA, &H5D.

For a normal program execution, remember to comment or erase instructions $sim, wait 1 and Portb = not CRC later.

Dallas Semiconductors offers more details on the 8-bit CRC check. The Appendix refers to useful links.

4.9 SPI Interface

The SPI interface uses three lines for serial communication.

In addition to many memory devices compatible with SPI, all well-known manufacturers offer analog-to-digital (ADC) and digital-to-analog converters (DAC), RTC devices, and others.

As shown in Figure 83, the microcontroller sends the serial data via its MOSI (*Master Out Slave In*) line to input SI of the peripheral device. The peripheral device sends its data via output SO to the MISO (*Master In Slave Out*) line. The data exchange is clocked by SCK. The microcontroller generates this clock signal.

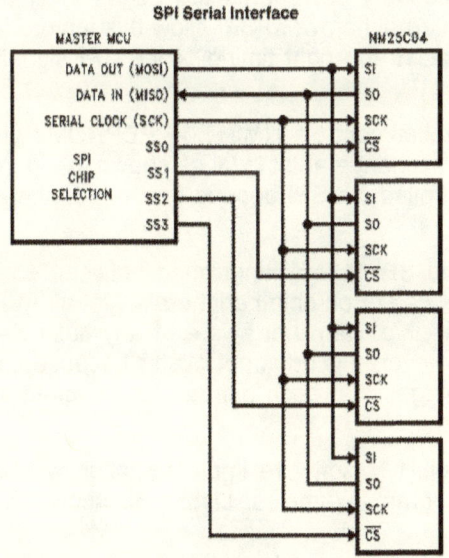

Figure 83   SPI Interface

The Chip Select signals SS0 to SS3 activate the peripheral device to be accessed.

Figure 84 shows the timing for data exchange between the microcontroller and EEPROM NM25C04 via the SPI interface.

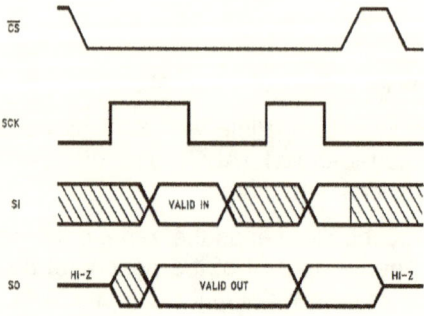

Figure 84   SPI Timing (NM25C04)

Some modifications are necessary as regards the edges of clock SCK. Microcontrollers with internal hardware SPI allow this feature to be configured. In all other cases the right timing of the SPI signals needs to be programmed.

In Figure 83 the microcontroller controls NM25C04 EEPROMs via SPI. SPI describes the data exchange at bit level as shown in Figure 84. The functions to be controlled via SPI depend on the peripheral device used.

As to the configuration of the SPI, it is distinguished between software implementation and the use of on-chip peripherals. Not all AVR microcontrollers offer SPI on-chip. I am not aware of any 8051 derivative with SPI on-chip. However, this does not matter because the software SPI has the advantages that each pin can be assigned to that kind of digital I/O.

For reasons of a better flexibility, software implementation will be made use of in the next program examples. Listing 43 shows the output of one byte to the SPI.

The AVR microcontroller was used in all of these examples; if the I/O ports used are modified, however, the programs will also work with BASCOM-8051.

PortA is used for the SPI lines. Write instruction `Config Spi = ...` in one line!

```
' AVR SPI

Dim X As Byte
X = &HAA

Config Spi = Soft , Din = Porta.0 , Dout = Porta.1 , Ss = Porta.2 , Clock = Porta.3

Spiinit

Spiout X , 1

nop

End
```

Listing 43   SPI Byte Output (SPI.BAS)

Program SPI.BAS shows the declaration of a byte variable X and the initialization of this variable using a value of &HAA.

After the configuration and initialization of the SPI interface, instruction `Spiout X, 1` sends the data byte X. The following `nop` was included for setting a break point during simulation.

The SPI interface works like an 8-bit shift register. The byte to be sent is saved in a register and will be shifted bit-by-bit to pin MOSI (master-out slave-in). The free positions are filled with bits received from pin MISO (master-in slave-out). After eight clocks the whole byte is sent and the register contains the complete byte received.

To send and receive bytes at the same time, function `Spimove()` should be used according to Listing 44. This function is available for BASCOM-AVR only.

If the same function is desired to be used for BASCOM-8051, it will be best to program it in BASIC.

```
' SPIMOVE by AVR

Dim A As Byte

A = &HAA

Config Spi = Soft , Din = Porta.0 , Dout = Porta.1 , Ss = Porta.2 ,
Clock = Porta.3

Spiinit

A = Spimove(a)

nop

End
```

Listing 44   SPI Read and Write at the Same Time (SPI1.BAS)

The AVR microcontrollers have on-chip SPI. If this internal peripheral is intended to be used, the fixed pin allocation must be taken into account. Figure 85 shows the settings in menu **Options>Compiler> I2C, SPI, 1WIRE**. Line `Config Spi = ...` is not needed in the program source.

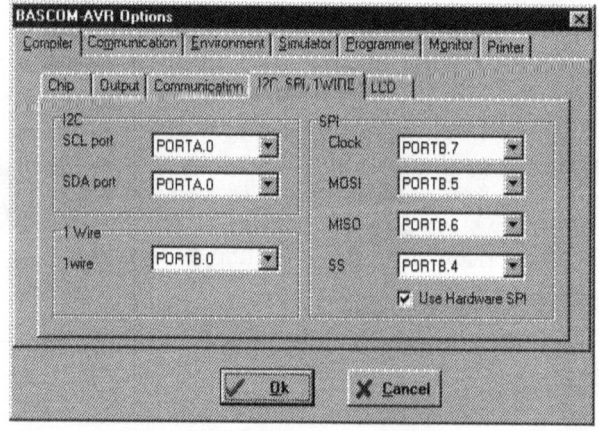

Figure 85   Configuration of On-Chip SPI

Listing 45 shows the source for data exchange via SPI using the on-chip peripheral. It seems there is nothing changed.

If the resulting assembler code in the AVR Studio is inspected, great differences will be found in the length of the assembler code.

```
Dim X As Byte
X = &HAA

Spiinit

Spiout X , 1

nop

End
```

Listing 45   Data Exchange via On-Chip SPI (SPI4.BAS)

The SPI Control Register organizes the whole SPI data exchange. AVR Studio can show the initialization by instruction Spiinit in detail. Figure 86 shows the initialization of the SPI Control Registers after running instruction Spiinit.

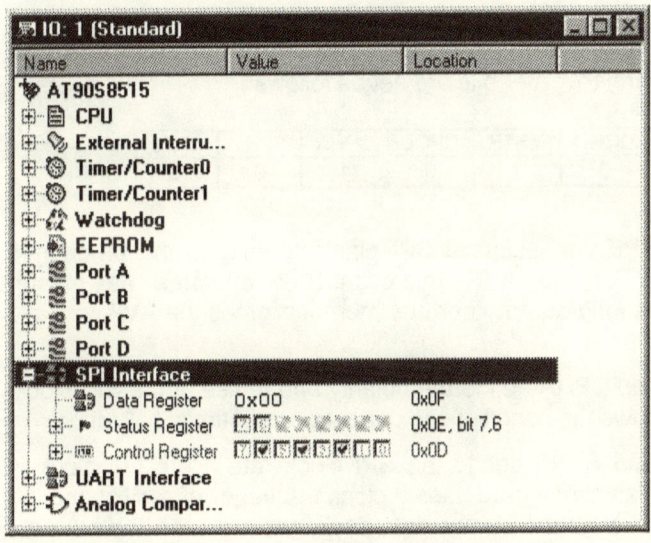

Figure 86   Initialization of the SPI Control Register

For the initialization in the BASCOM-AVR simulator, see the IO register contents as shown in Figure 87.

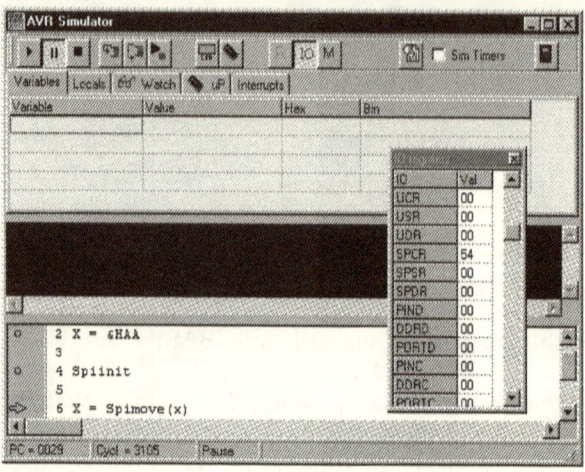

Figure 87
Initialization of the SPI Control Register

The SPI Control Register bits are set as follows:

SPIE	SPE	DORD	MSTR	CPOL	CPHA	SPR1	SPR0	
0	1	0	1	0	1	0	0	SPCR

Setting bit SPE connects the SPI pins internally to the predefined pins of PortB. The AVR microcontroller operates as master (MSTR=1) as long as no other bus member forces the AVR to slave via SS line.

Bits CPOL and CPHA define the polarity and phase of the SPI clock. Figure 88 shows the conditions in dependence of the initialization.

Bits SPR1 and SPR0 define the SPI clock rate. Here the clock is CK/4. The evaluation board uses a clock frequency of 4 MHz and the SPI clock is 1 MHz. As always data packages are sent, the net data rate of 1 Mbit/s is not reached.

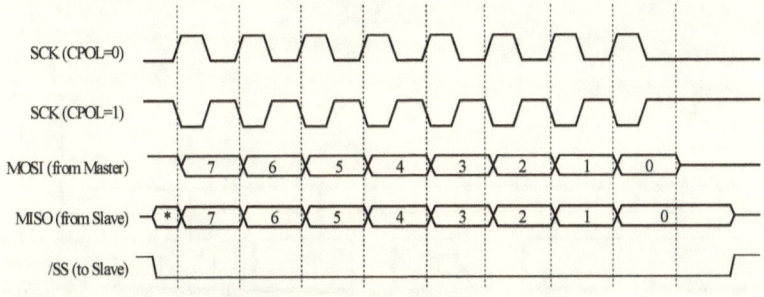

Figure 88   SPI Data Exchange with CPHA=1

Initialization by instruction Spiinit can be changed any time by direct manipulation of the SPI Control Register.

4.10 I^2C Bus

The I^2C Bus was developed for data exchange between different devices like EEPROMs, RAMs, AD- and DA-converters, RTCs and microcontrollers in a network environment.

Figure 89 shows the connections required in a typical I^2C bus network. Lines SDA and SCL, connected via pull-up resistors to the supply voltage V_{CC}, connect all members of the network. An I^2C bus network can connect different masters to different slaves (multi-master system). The I^2C protocol is responsible for addressing the individual nodes.

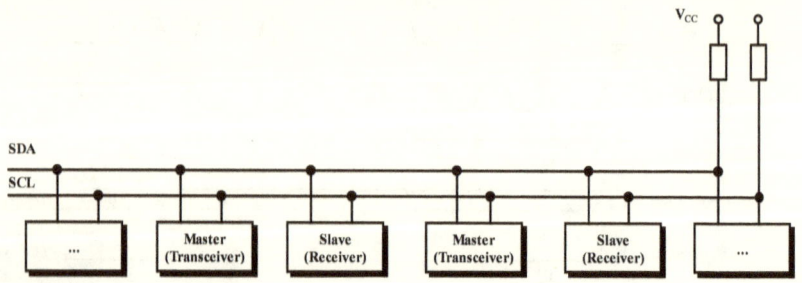

Figure 89   I²C Bus Network

The resulting peripheral functions are device-specific. In addition to many EEPROMs and RAMs from different manufacturers, there are many specialized I²C bus devices:

- I/O expander devices
- LCD and LED driver devices
- video controller
- PAL/NTSC TV processors
- TV and VTR stereo/dual sound processors with integrated filters
- Hi-Fi stereo audio processor interface for color decoder
- YUV/RGB switches
- programmable modulators for negative-video modulation and FM sound
- satellite sound receiver
- programmable RF modulators
- BTSC stereo/SAP decoder and audio processor
- 1.3 and 2.5 GHz bi-directional synthesizer
- 1.4 GHz multimedia synthesizer

Before we deal with the first I²C bus program example, I would like to explain some frequently used terms.

Term	Explanation
WORD	8 data bits
PAGE	16 consecutive memory locations
PAGE BLOCK	2048 bits organized in 16 pages
MASTER	any I^{2}C device controlling data exchange (a microcontroller, for example)
SLAVE	controlled I^{2}C device
TRANSMITTER	I^{2}C device sending data to the I^{2}C bus (master or slave)
RECEIVER	I^{2}C device receiving data from I^{2}C bus (master or slave)
TRANSCEIVER	I^{2}C device containing transmitter and receiver

In the I^{2}C bus program example, one I^{2}C bus EEPROM is connected to two I/O pins of the AVR microcontroller. Due to the required read and write operations, memory devices are well suited for describing of I^{2}C bus operations.

The slave address of each I^{2}C bus device contains the Device Type Identifier. The used EEPROM of the NM24Cxx family has the following slave address. The Device Type Identifier is here &B1010.

1	0	1	0	A2	A1	A0	R/W

A further part of the slave address is the device address. To define a device address, address pins A2, A1 and A0 must be connected to V$_{CC}$ or GND. The next table shows the active address pins of the NM24Cxx family.

Device	A2	A1	A0	Memory
NM24C02	addr	addr	addr	2 K
NM24C04	addr	addr	x	4 K
NM24C08	addr	x	x	8 K
NM24C16	x	x	x	16 K

As shown in the table, one I²C network can address max. 16 Kbit (16384 bits) of memory. It does not matter whether one NM24C16 or eight NM24C02 or other configurations are used.

For addressing an EEPROM, there are two different addressing levels:

1. Hardware configuration by pins A2, A1 and A0 (Device Address Pins) with pull-up or pull-down resistors. All unused pins (marked with x in the table) must be connected to GND.
2. Software addressing of the used memory segment (Page Block) within the memory of the used device.

For addressing the memory in EEPROM the respective command must provide the following information:

[DEVICE TYPE]
[DEVICE ADDRESS]
[PAGE BLOCK ADDRESS]
[BYTE ADDRESS]

In the program example, the EEPROM NM24C16 is used. Because of its 16 Kbit memory there is no hardware configuration possible. Pins A2, A1 and A0 must be connected to GND.

Bits A2, A1 and A0 of the slave address point to an internal memory segment (PAGE BLOCK). The LSB of the slave address defines writing (HI) or reading (Lo).

Byte Write (write one byte to any memory location) and Random Read (read one byte from any memory location) are two basic functions for data exchange via the I²C bus. To make access to the memory in an EEPROM more effective, there are further possibilities for access like Page Write, Current Address Read and Sequential Read.

The program example focuses on the basic functions. Using the knowledge acquired, it should be no problem to complete the special functions. Figure 90 shows the bit sequences for the Byte Write and Random Read operations.

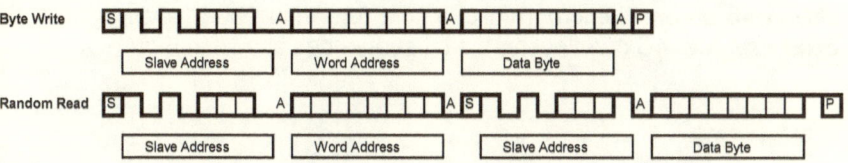

Figure 90   Byte Write and Random Read Operations

Every command begins with a start condition (S). The start condition is defined by a falling edge on SDA during SCL = Hi. Each I^{2}C bus device permanently detects the levels on the SDA and SCL lines to find a valid start condition. If no valid start condition is found, no devices will answer.

The first byte after a start condition is the slave address showing a write access to the addressed memory segment. The transmitting device releases the I^{2}C bus after eight transmitted bits. During the ninth clock the receiver forces line SDA to Lo to ACKnowledge (A) the eight bits received. This acknowledge mechanism is a software agreement for a successful data exchange.

The second byte sent addresses the memory location inside the addressed memory segment for a following read or write operation. The last transmitted bits are followed by the acknowledge check.

In a Byte Write operation the data byte is sent as the third byte. The last transmitted bits are again followed by the acknowledge check.

In a Random Read operation following the check for acknowledge, a new start condition must be sent. The first byte after this new start condition is a slave address and an indicated read access to the memory location addressed before. After the last bits have been sent, the check for acknowledge is carried out again to read the addressed EEPROM cell.

During a read access the I^{2}C bus slave sends eight data bits and then checks the acknowledge from the master. If acknowledge is detected and no stop condition is sent from the master, the slave will send further data. If acknowledge is not detected, the slave will stop sending data and waits for a stop condition to return to standby.

Each data exchange ends with a stop condition (P). The stop condition is defined by a rising edge on line SDA during SCL = Hi. The stop condition additionally switches the EEPROMs of the NM24Cxx family to the current saving standby mode.

Listing 46 shows the program example for writing and reading the EEPROM NM24Cxx reflecting the bit sequences shown in Figure 90.

```
' I2C for AVR

Const Device_id = &HA           ' Device ID for NM24Cxx
Const Page_addr = 1             ' used Page
Const Word_addr = 0             ' used memory location
Const Ee_data = &HA5            ' used data byte

Dim Slave_wa As Byte            ' Slave Write Address
Dim Slave_ra As Byte            ' Slave Read Address
Dim Temp As Byte

Config Scl = Porta.0            ' PortA.0 is SCL
Config Sda = Porta.1            ' Porta.1 is SDA

Config Portb = Output           ' Portb is output
Portb = 255  ' LEDs off

Slave_wa = Device_id            ' Calculation of Slave Address
Shift Slave_wa , Left , 4

Temp = Page_addr
Shift Temp , Left

Slave_wa = Slave_wa Or Temp
Slave_ra = Slave_wa Or 1

I2cstart                        ' I2C Write Sequence
I2cwbyte Slave_wa
I2cwbyte Word_addr
I2cwbyte Ee_data
I2cstop

Waitms 10                       ' Wait for end of program cycle

I2cstart                        ' I2C Read Sequence
I2cwbyte Slave_wa
I2cwbyte Word_addr
I2cstart
I2cwbyte Slave_ra
I2crbyte Temp , Nack
I2cstop

Portb = Temp ' Display read EEPROM data

End
```

Listing 46   Access to I²C EEPROM NM24C16 by AVR (IIC.BAS)

To simplify the procedure, some parameters were defined as constants:

- Device Identifier for all EEPROMs of the NM24Cxx family `device_id = &HA`
- Memory access to page 1 address 0 by `page_addr = 1` and `word_addr = 0` (random)
- Data byte for writing `ee_data = $A5` (random)

To change the address and/or data byte, only the respective constants are to be changed.

Pins PortA.0 and PortA.1 serve as I²C bus lines SCL and SDA. PortB serves as output. Driving the connected LEDs (of the evaluation board) it displays the data byte read back from EEPROM.

Thereafter, the addresses `Slave_wa` (slave write address) and `Slave_ra` (slave read address) are calculated.

As in Figure 90, the single I²C bus instructions will follow. Since BASCOM-AVR hides the details of implementation, the programming of these sequences poses no problem.

The last instruction before end of program writes the data byte read back from EEPROm to PortB. If the program works properly, bit pattern &HA5 will appear at PortB.

BASCOM-AVR takes care of the right I²C bus timing for all possible clock frequencies.

4.11 Scalable Network Protocol S.N.A.P

For the networking of computers and microcontrollers, numerous protocols are known and in use today. These protocols guarantee an errorless communication between different network nodes. To implement such network protocols, resources are often required which are not available for small microcontrollers.

Therefore a simple but scalable network protocol ready for implementation in existing microcontroller applications is desirable.

High Tech Horizon, a Swedish company [http://www.hth.com], developed for their powerline modems PLM-24 such a simple network for protocols which is suitable for both small microcontrollers and larger systems.

The scalable network protocol S.N.A.P. (Scalable Node Address Protocol) is the result of this development.

4.11.1 S.N.A.P. Features

S.N.A.P. has many features which are listed and explained below:

- Easy to learn, use and implement.
- Free and open network protocol.
- Free development tools available.
- Scaleable binary protocol with small overhead.
- Requires minimal MCU resources to implement.
- Up to 16.7 million node addresses.
- Up to 24 protocol specific flags.
- Optional ACK/NAK request.
- Optional command mode.
- 8 different error detecting methods (Checksum, CRC, FEC etc.).
- Can be used in master/slave and/or peer-to-peer.
- Supports broadcast messages.
- Media independent (power line, RF, TP, IR etc.).
- Works with simplex, half, full duplex links.
- Header is scalable from 3-12 bytes.
- User specified number of preamble bytes (0-n).
- Works in synchronous and asynchronous communication.
- Works with HTH's free PLM-24 < > TCP/IP Gateway software.

Don't be afraid of this extensive list. It is typical of a scalable solution to take precautions for larger systems in the general approach. For implementation in a small system, a minimum approach is absolutely sufficiently.

4.11.2 Description of S.N.A.P. Protocol

Communication between network nodes is in the form of data packages. These data packages can have different lengths. The total length will be determined by the number of address and data bytes, the error detection method and some specific bytes.

The Header Definition Bytes, HDB2 and HDB1, determine the structure of the data package (telegram) and its length. Each telegram can have an uncertain number of preamble bytes before the synchronization byte. The preamble byte must differ from the synchronization byte.

The following example shows a small S.N.A.P. package with CRC16 error detection:

PRE	...	SYNC	HDB2	HDB1	DAB1	SAB1	DB1	CRC2	CRC1

It means:

Name	Description
PRE	Preamble
SYNC	Synchronization
HDB2	Header Definition Byte 2
HDB1	Header Definition Byte 1
DAB1	Receiver address
SAB1	Transmitter address
DB1	Data byte
CRC2	Most significant byte of CRC16
CRC1	Least significant byte of CRC16

Without the optional preamble bytes the whole data package is eight bytes long. The bytes are right positioned with its LSB (least significant bit; bit7...bit0).

4.11.2.1 Synchronization Byte (SYNC)

Byte SYNC is predefined and marks the beginning of each data package.

bit	7	6	5	4	3	2	1	0	HEX	DEC
	0	1	0	1	0	1	0	0	54	84

4.11.2.2 Header Definition Bytes (HDB2 and HDB1)

Following byte SYNC are the Header Definition Bytes HDB2 and HDB1 which determine the structure of the telegram.

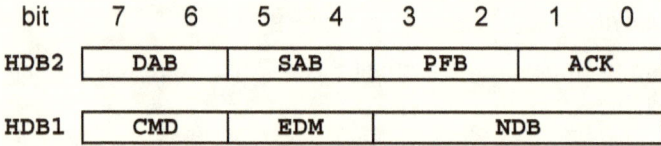

The bits in HDB2 and HDB1 have the following meaning:

Name	Description
DAB	Number of bytes for destination address
SAB	Number of bytes for source address
PFB	Number of bytes for protocol specific flags
ACK	ACK/NAK bits
CMD	Command bit
EDM	Error detection method
NDB	Number of data bytes

The following conditions apply to Header Definition Byte HDB2:

DAB		Definition
0	0	Destination address 0 Byte
0	1	Destination address 1 Byte
1	0	Destination address 2 Byte
1	1	Destination address 3 Byte

SAB		Definition
0	0	Source address 0 Byte
0	1	Source address 1 Byte
1	0	Source address 2 Byte
1	1	Source address 3 Byte

PF		Definition
0	0	Protocol specific flags 0 Byte
0	1	Protocol specific flags 1 Byte
1	0	Protocol specific flags 2 Byte
1	1	Protocol specific flags 3 Byte

The flag bytes are reserved for the time being, but not defined yet. They are planned for further enhancements of the S.N.A.P. protocol.

ACK		Definition
0	0	No acknowledge
0	1	Transmitter requests for acknowledge
1	0	Receiver sends back ACK
1	1	Receiver sends back NAK

The following conditions apply to Header Definition Byte HDB1:

CMD	Definition
0	No command mode
1	Command mode (DB1 contains command)

A network node in the command mode offers more flexibility. If the command bit is set (CMD=1) then the data byte DB1 contains a command. Different commands are possible due to the byte format 256.

It is dependent on the error detection method how safely a communication link works. The 16-bit CRC is a preferred method in this area.

EDM			Definition
0	0	0	No error detection
0	0	1	Repeat three times
0	1	0	8-bit check sum
0	1	1	8-bit CRC-CCITT
1	0	0	16-bit CRC-CCITT
1	0	1	32-bit CRC-CCITT
1	1	0	Error correction
1	1	1	Spec. error detection

	ND	M		Definition
0	0	0	0	0 Byte
0	0	0	1	1 Byte
0	0	1	0	2 Byte
0	0	1	1	3 Byte
0	1	0	0	4 Byte
0	1	0	1	5 Byte
0	1	1	0	6 Byte
0	1	1	1	7 Byte
1	0	0	0	8 Byte
1	0	0	1	16 Byte
1	0	1	0	32 Byte
1	0	1	1	64 Byte
1	1	0	0	128 Byte
1	1	0	1	256 Byte
1	1	1	0	512 Byte
1	1	1	1	Spec. number

4.11.3 S.N.A.P. Monitor

A simple program example serves to explain the implementation and use of the S.N.A.P. protocol for data exchange in a master/slave system by means of a S.N.A.P. monitor.

To simplify this example, acknowledge and error detection shall be excluded for the time being. The telegram looks as follows and is described by the comment lines at the beginning of the program.

| SYNC | HDB2 | HDB1 | DAB1 | SAB1 | DB1 |

The addresses are reduced to one byte. Only one byte will be transferred. Header Definition Bytes HDB2 and HDB1 contain the following values:

The program waits for a telegram to be sent, and analyzes it as shown in Figure 91.

bit	7	6	5	4	3	2	1	0	HEX	DEC
HDB2	DAB		SAB		PFB		ACK			
	0	1	0	1	0	0	0	0	50	80
HDB1	CMD		EDM		NDB					
	0	0	0	0	0	0	0	1	00	01

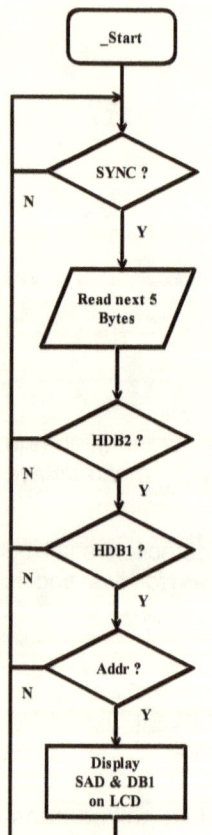

After the start of the program it will wait for a SYNC byte.

If a SYNC byte was detected, the program reads five data bytes as defined in the Header Definition bytes, and evaluates these bytes.

It depends on the evaluation of the Header Definition bytes whether the telegrams can be understood or not.

If the telegram can be evaluated, then the program checks the receiver address.

Only if all information is correct, the program reads from byte SADR the address of the transmitter and from byte DB1 the respective data byte. Both data will be displayed on LCD in the example.

Listing 47 shows the details of the S.N.A.P. application. The area of the associated function is specially marked.

Figure 91
Program structure
SNAP-MON

```
' -----[ Title ]----------------------------------------
'
' File....: SNAP-MON.BAS
' Purpose.: SNAP Monitor with 16 x 1 LCD
' Author..: Claus Kuehnel  (based on SNAP-024.BAS from HTH)
' Version.: 1.00
' MCU.....: AT90S8515
' Started.. 991211
' Updated.:
'
' -----[ Program Description ]----------------------------
'
' This program shows how to implement the S.N.A.P protocol
' in BASCOM and is an simple example to receive data from
' another node and display it on the LCD.
' This example uses no error detection method.
'
' If the node is addressed by another node (PC or another
' MCU) it shows what node sent the packet and the value of
' DB1 in hexadecimal on the LCD. In the example below node
' with address &H05 is sending &HEF in the data byte.
'
'   16x1 LCD display
' +------------------+
' | FROM:05  DATA:EF |
' +------------------+
'
' The packet structure is defined in the received packets
' first two bytes (HDB2 and HDB1). The following packet
' structure is used.
'
' DD=01       - 1 Byte destination address
' SS=01       - 1 Byte source address
' PP=00       - No protocol specific flags
' AA=00       - Acknowledge is required
' C=0         - Command mode not supported
' EEE=000     - No error detection
' NNNN=0001   - 1 Byte data
'
'
' -----[ Aliases ]----------------------------------------
'
Spkr Alias Portb.1                  ' Speaker output pin
'
' -----[ Constants ]--------------------------------------
'
Const Sync = &B01010100             ' Synchronisation byte
Const Hdb2_ = &B01010000            ' 1-Byte Addr., No Flags, No ACK
Const Hdb1_ = &B00000001  ' No Error Detection, 1 Data Byte
Const Myaddress = &H04              ' Address for this node (1-255)

' -----[ Variables ]--------------------------------------
'
Dim Temp1 As Byte                   ' Temporary Variable
Dim Temp2 As Byte                   ' Temporary Variable
Dim Hdb2 As Byte                    ' Header Definition Byte 2
Dim Hdb1 As Byte                    ' Header Definition Byte 1
Dim Db1 As Byte                     ' Packet data
Dim Sab1 As Byte                    ' What node sent this packet
```

```
Dim Dab1 As Byte                         ' What node should have this paket

' -----[ Initialization ]--------------------------------
' Configure LCD display
Config Lcd = 16 * 1

Cls                                      ' Clear the LCD display
Cursor Off Noblink                       ' Hide cursor

'------[ Program ]--------------------------------------
'

_start:
    Temp1 = Waitkey()                    ' Wait for data on serialport

    ' If received data is a SYNC byte read next five bytes
    ' from master, if not return to start
    If Temp1 <> Sync Then
        Goto _start
    Else
        ' Get packet in binary mode
        Inputbin Hdb2 , Hdb1 , Dab1 , Sab1 , Db1

        ' Packet header check routine
        '
        ' Check HDB2 to see if MCU are capable to use the
        ' packet structure, if not goto Start
        If Hdb2 <> Hdb2_ Then Goto _start

        ' Check HDB1 to see if MCU are capable to use the
        ' packet structure, if not goto Start
        If Hdb1 <> Hdb1_ Then Goto _start

        ' Address check routine
        '
        ' Check if this is the node addressed,
        ' if not goto Start
        If Dab1 <> Myaddress Then Goto _start

        ' Associated function (place it between +++ lines)
' ++++++++++++++++++++++++++++++++++++++++++++++++++++++
        Cls
        Lcd "FROM:"
        Lcd Hex(sab1)
        Lcd " "
        Lcd "DATA:"
        Lcd Hex(db1)
        ' Beep to alert new message
        Sound Spkr , 10000 , 10          'BEEP
' ++++++++++++++++++++++++++++++++++++++++++++++++++++++
```

```
        ' Done, go back to Start and wait for a new packet
        Goto _start
    End If
End
```

Listing 47   AVR S.N.A.P. Monitor (SNAP-MON.BAS)

To modify the program for BASCOM-8051, the port line for the speaker (`Spkr Alias Portb.1` to `Spkr Alias P3.1`, for example) and the display instructions `lcd hex(var)` to `lcdhex var` need to be changed.

4.11.4 Digital I/O

In the next program example, a network node converts serially received data to digital I/O.

For an errorless data exchange the 16-bit-CRC serves the detection of transmission errors. ACK or NAK reports the result of data transmission back to the transmitter.

If the transmission was not correct, the transmitter can repeat the transmission of that data package.

The following telegram structure shall be used in the example:

SYNC	HDB2	HDB1	DAB1	SAB1	DB2	DB1	CRC2	CRC1

Here, too, the addresses are reduced to one byte each. Two data bytes, DB2 and DB1, and two CRC bytes, CRC2 and CRC1, will be transmitted.

Header Definition Bytes HDB2 and HDB1 contain the following values:

bit	7	6	5	4	3	2	1	0	HEX	DEC
HDB2	DAB		SAB		PFB		ACK			
	0	1	0	1	0	0	0	1	51	81
HDB1	CMD		EDM		NDB					
	0	1	0	0	0	0	1	0	42	66

Listing 48 shows the program that waits for receiving a telegram and evaluates it according to Figure 92.

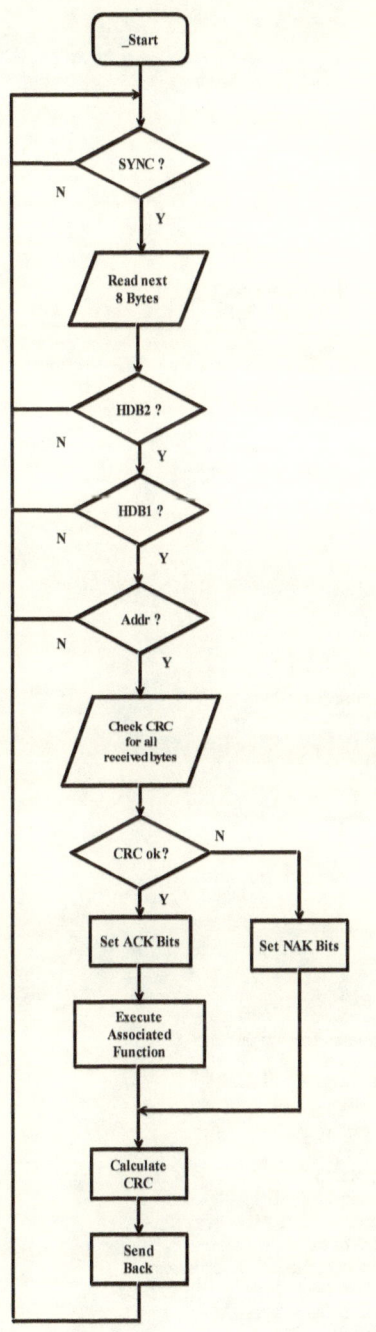

Figure 92
Program Structure SNAP-IO.BS2

After the start of the program it will wait for receiving a SYNC byte.

When a SYNC byte is detected, the program reads eight data bytes as defined in the Header Definition bytes, and evaluates these bytes.

The evaluation of the Header Definition bytes is the same as in the last program example.

If the destination address was correct, the 16-bit CRC of all received bytes will be calculated.

If the CRC is correct, the acknowledge bits in HDB2 are set to 10_B and the associated function will be executed.

In Listing 48 the area of the associated function is marked. In our example the bits of data byte DB1 are written to PortB and displayed by the connected LEDs.

A wrong CRC indicates a transmission error. In this case the program sets the acknowledge bits to 11_B and the associated function will <u>not</u> be executed.

Thereafter, the node sends a telegram of the same structure to the sender. The sender can evaluate this response now.

```
' -----[ Title ]-------------------------------------------
'
' File......: SNAP-IO.BAS
' Purpose...: Turns LEDs on and off
' Author....: Christer Johansson
' Version...: 1.01
' Started...: 980503
' Updated...: 980918
' Modified..: 991229 by Claus Kuehnel
'
' -----[ Program Description ]-----------------------------
'
' This program shows how to implement the S.N.A.P protocol
' in BASCOM-AVR and is an simple example to turn LEDs ON or
' OFF.
' This example uses 16-bit CRC-CCITT as error detection
' method which gives secure data transfer.
'
' The packet structure is defined in the received packets
' first two bytes (HDB2 and HDB1). The following packet
' structure is used.
'
' DD=01      - 1 Byte destination address
' SS=01      - 1 Byte source address
' PP=00      - No protocol specific flags
' AA=01      - Acknowledge is required
' D=0        - No Command Mode
' EEE=100    - 16-bit CRC-CCITT
' NNNN=0010  - 2 Byte data
'
' Overview of header definition bytes (HDB2 and HDB1)
'
'        HDB2              HDB1
' +-----------------+-----------------+
' | D D S S P P A A | D E E E N N N N |
' +-----------------+-----------------+
'
'
' -----[ Constants ]---------------------------------------

Const Preamble_ = &B01010101        ' Preamble byte
Const Sync_     = &B01010100                ' Synchronisation byte
Const Crcpoly   = &H1021                    ' CRC-CCITT
Const Hdb2_     = &H51
Const Hdb1_     = &H42
Const Myaddress = 123               ' Address for this node (1-255)

' -----[ Variables ]---------------------------------------

Dim Preamble As Byte                ' Preamble byte
Dim Sync As Byte                    ' Sync byte
Dim Crc As Word                     ' CRC Word
Dim Hdb1 As Byte                    ' Header Definition Byte 1
Dim Hdb2 As Byte                    ' Header Definition Byte 2
Dim Dab1 As Byte        ' What node should have this paket
Dim Sab1 As Byte                    ' What node sent this packet
Dim Db1 As Byte                     ' Packet Data Byte 1
Dim Db2 As Byte                     ' Packet Data Byte 2
Dim Crc2 As Byte                    ' Packet CRC Hi_Byte
Dim Crc1 As Byte                    ' Packet CRC Lo_Byte
Dim Temp1 As Byte                   ' Temporary Variable
```

```
Dim Temp2 As Byte                       ' Temporary Variable
Dim Tmpw1 As Word
Dim Tmpw2 As Word

' -----[ Initialization ]--------------------------------
'
Config Portb = Output                   ' Portb is output
Portb = &HFF

Preamble = Preamble_
Sync = Sync_

Db1 = 0                                 ' Clear Data variable
Db2 = 0

'------[ Program ]---------------------------------------
'
_start:
    Temp1 = Waitkey()                   ' Wait for data on serialport
    ' If received data is a SYNC byte read next eight bytes
    ' from master, if not return to start
    If Temp1 <> Sync Then
        Goto _start
    Else
        ' Get packet in binary mode
Inputbin Hdb2 , Hdb1 , Dab1 , Sab1 , Db2 , Db1 , Crc2 , Crc1

        ' Packet header check routine
        '
        ' Check HDB2 to see if MCU are capable to use the
        ' packet structure, if not goto Start
        If Hdb2 <> Hdb2_ Then Goto _start

        ' Check HDB1 to see if MCu are capable to use the
        ' packet structure, if not goto Start
        If Hdb1 <> Hdb1_ Then Goto _start

        ' Address check routine
        '
        ' Check if this is the node addressed,
        ' if not goto Start
        If Dab1 <> Myaddress Then Goto _start

        ' Check CRC for all the received bytes
        Gosub Check_crc

        ' Check if there was any CRC errors, if so send NAK
        If Crc <> 0 Then Goto Nak

        ' No CRC errors in packet so check what to do.
```

```
' Associated Function (place it between +++ lines)
'+++++++++++++++++++++++++++++++++++++++++++++++++++++
       Portb = Db1
'+++++++++++++++++++++++++++++++++++++++++++++++++++++

Ack_:
         ' Send ACK (i.e tell master that packet was OK)
         ' Set ACKs bit in HDB2 (xxxxxx10)
         Hdb2 = Hdb2 Or &B00000010
         Hdb2 = Hdb2 And &B11111110
         Goto Send

Nak:
         ' Send NAK (i.e tell master that packet was bad)
         ' Set ACK bits in HDB2 (xxxxxx11)
         Hdb2 = Hdb2 Or &B00000011
         Goto Send

Send:
         ' Swap SAB1 <-> DAB1 address bytes
         Temp2 = Sab1
         Sab1 = Dab1
         Dab1 = Temp2

         ' Clear CRC variable
         Crc = 0

         ' Put HDB2 in variable Tmp_Byte1
         Temp1 = Hdb2
         ' Calculate CRC
         Gosub Calc_crc

         ' Put HDB1 in variable Tmp_Byte1
         Temp1 = Hdb1
         ' Calculate CRC
         Gosub Calc_crc

         ' Put DAB1 in variable Tmp_Byte1
         Temp1 = Dab1
         ' Calculate CRC
         Gosub Calc_crc

         ' Put SAB1 in variable Tmp_Byte1
         Temp1 = Sab1
         ' Calculate CRC
         Gosub Calc_crc

         ' Put Data in variable Tmp_Byte1
         Temp1 = Db2
         ' Calculate CRC
         Gosub Calc_crc

         ' Put Data in variable Tmp_Byte1
         Temp1 = Db1
         ' Calculate CRC
         Gosub Calc_crc

         ' Move calculated Hi_CRC value to outgoing packet
         Crc2 = High(crc)
         ' Move calculated Lo_CRC value to outgoing packet
```

```
      Crc1 = Low(crc)

     ' Send packet to master,
     ' including the preamble and SYNC byte
     Print Chr(preamble) ; chr(sync) ;
     Print Chr(hdb2) ; Chr(hdb1) ;
     Print chr(dab1) ; chr(sab1) ;
     Print chr(db2) ; chr(db1) ;
     Print chr(crc2) ; chr(crc1) ;

     ' Give AVR time to shift out all bits
     ' before setting to Rx
     Waitms 50

     ' Done, go back to Start and wait for a new packet
     Goto _start
   End If

' -----[ Subroutines ]------------------------------------
'
'Soubroutine for checking all received bytes in packet
Check_crc:
     Crc = 0
     Temp1 = Hdb2
     Gosub Calc_crc
     Temp1 = Hdb1
     Gosub Calc_crc
     Temp1 = Dab1
     Gosub Calc_crc
     Temp1 = Sab1
     Gosub Calc_crc
     Temp1 = Db2
     Gosub Calc_crc
     Temp1 = Db1
     Gosub Calc_crc
     Temp1 = Crc2
     Gosub Calc_crc
     Temp1 = Crc1
     Gosub Calc_crc
     Return
```

```
' Subroutine for calculating CRC value in variable Tmp_Byte1
Calc_crc:
        Tmpw1 = Temp1 * 256
        Crc = Tmpw1 Xor Crc
        For Temp2 = 0 To 7
              If Crc.15 = 0 Then Goto Shift_only
              Tmpw2 = Crc * 2
              Crc = Tmpw2 Xor Crcpoly
              Goto Nxt
Shift_only:
              Crc = Crc * 2
Nxt:
        Next
        Return

' -----[ End ]--------------------------------------------
```

Listing 48   AVR S.N.A.P. I/O Node (SNAP-IO.BAS)

To modify the program for BASCOM-8051, the configuration line for PortB must be erased and the port must be changed (Portb to P3, for example).

For communication with such a network node, High Tech Horizon offers some free tools [http://www.hth.com/snap/].

Program SnapLab running on a PC generates telegrams and sends these to the network nodes. SnapLab receives the responses from the network and analyzes them.

For our simple example this means that a PC and S.N.A.P. I/O node are connected via RS-232. A real network (with more than two nodes) would use RS-485 or a power line modem from HTH, for example.

The first step is to set the communication parameters in SnapLab. Figure 93 shows the settings.

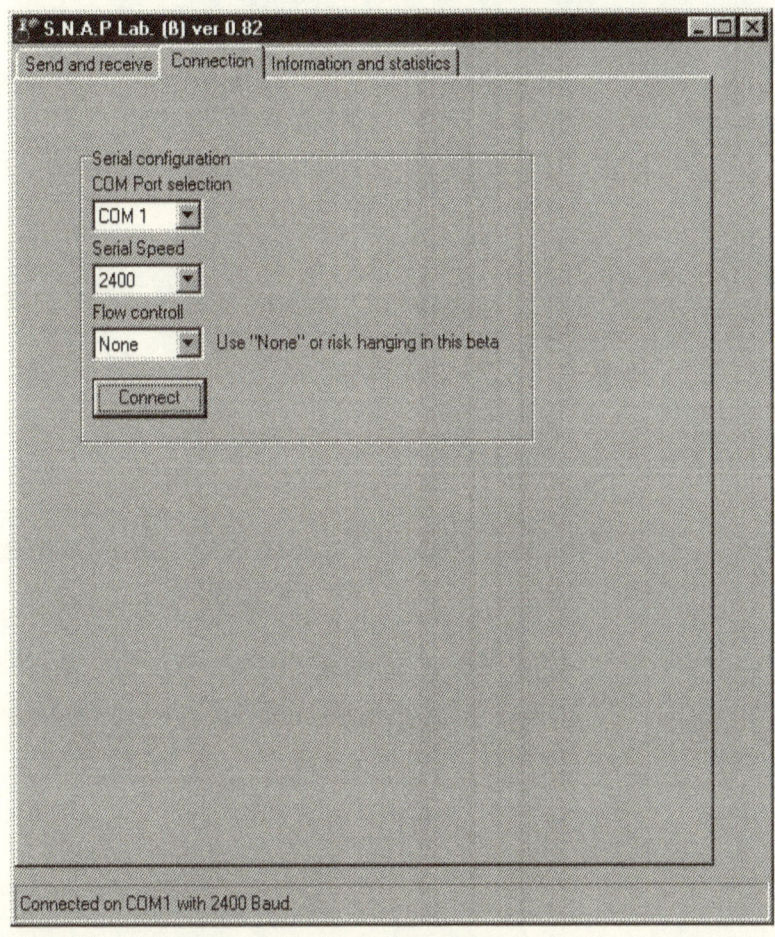

Figure 93   Configuration of the Serial Interface

After this first step, the telegrams can be built up. Figure 94 shows the respective window.

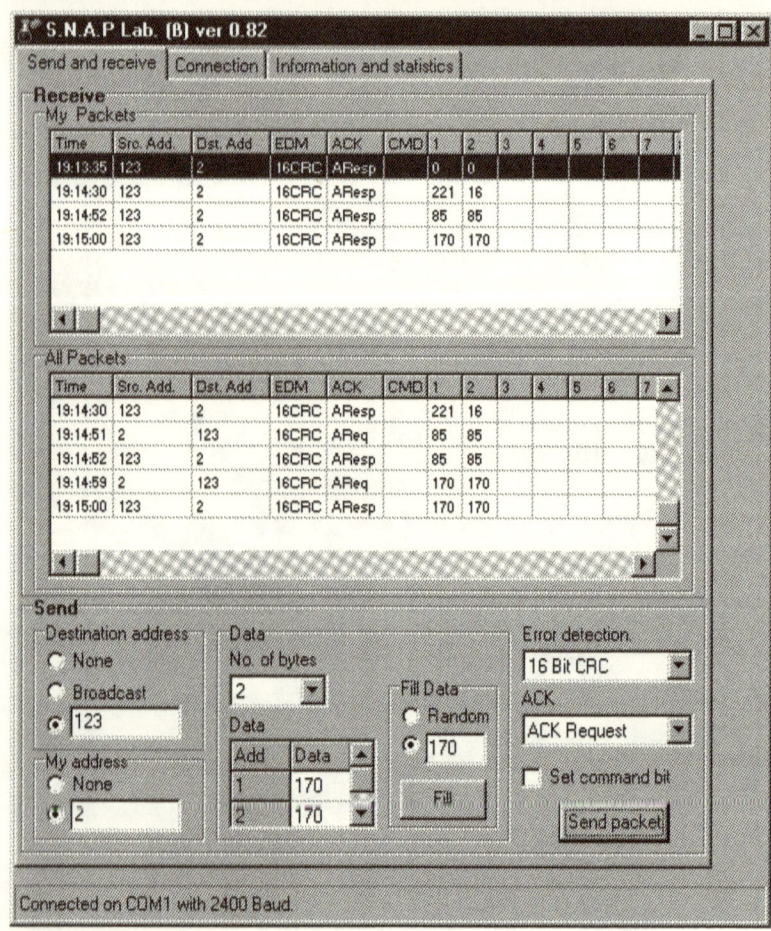

Figure 94 SnapLab - Transmit and Receive

The structure of the telegram is determined in the Send frame. In accordance with the program SNAP-IO.BAS the destination address for the S.N.A.P. I/O node is set to 123. The PC gets address 2.

A data package contains two data bytes initialized with 170 here. The 16-bit CRC is used for the detection of transmission errors. ACK or NAK report the result of data exchange to the transmitter.

There are two windows in the Receive frame. In the enter, all telegrams are listed with a time stamp. On top only the telegrams dedicated to the PC are listed.

These listings reveal, for example, that one telegram with DB1 = DB2 = &H170 was sent from PC to node 2 at 19:14:59. The response at

19:50:00 shows that the telegram was received without any error. Using the evaluation board for this test, the LEDs connected to PortB will show the related bit pattern &HAA.

The old DOS program RS232MON can be used to go deeper into the bits [http://www.ckuehnel.ch/download.htm]. Next, let's have a short look at the byte level.

The following telegram must be prepared to send the two data bytes &HAA and &H55 to the network node:

SYNC	HDB2	HDB1	DAD	SAD	DB2	DB1	CRC2	CRC1
84	81	66	123	1	170	85	243	96
&H54	&H51	&H42	&H7B	&H01	&HAA	&H55	&HF3	&H60

For the input of the data bytes to be sent to the network node, the number pad of the PC keyboard should be used.

For the input of data byte 84 for example, strike 0-8-4 keeping the Alt key pressed. The character will be sent when the Alt key is released.

After the start of program RS232MON and the configuration of the serial port, the telegram can be inputted as described. Figure 95 shows the input and the response from the network node. The byte sequences appear very cryptic.

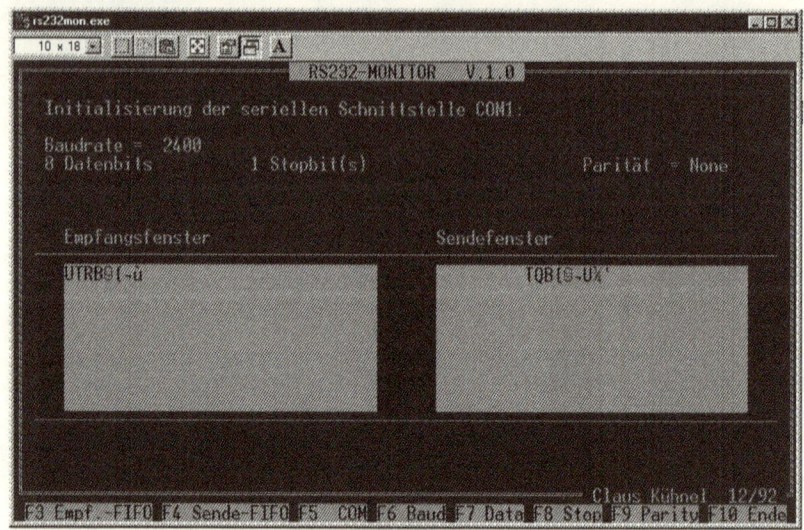

Figure 95   S.N.A.P. Telegram in RS232MON

A short look at the receive window (F3) shows that the telegram was transmitted without any errors (Figure 96).

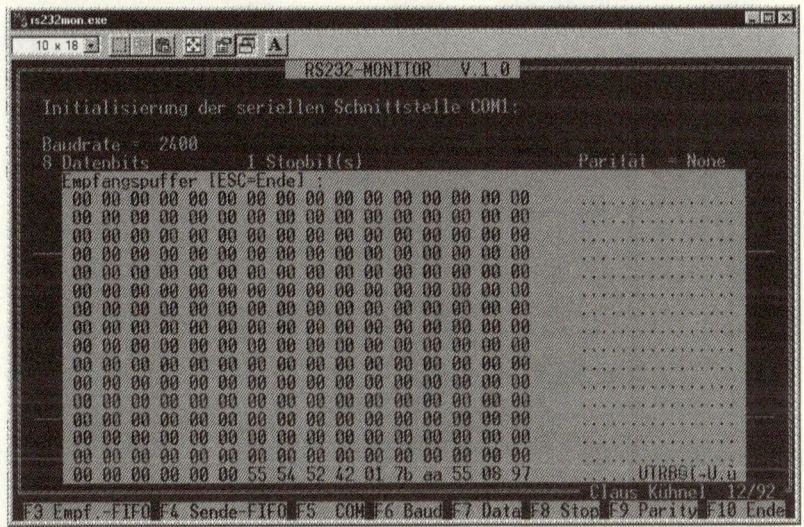

Figure 96   Hexdump of a Received S.N.A.P. Telegram

Due to the acknowledge bits (10) the Header Definition Byte HDB2 of the response is 82 (=&H52). The data bytes are unchanged. Source and destination addresses are swapped and the 16-bit CRC is changed.

SYNC	HDB2	HDB1	DAD	SAD	DB2	DB1	CRC2	CRC1
84	82	66	1	123	170	85	8	151
&H54	&H52	&H42	&H01	&H7B	&HAA	&H55	&H08	&H97

To simulate a transmission error, a wrong byte shall be typed for CRC1 in contrast to Figure 95, i.e. 97 instead of the correct value of 96 (Figure 97).

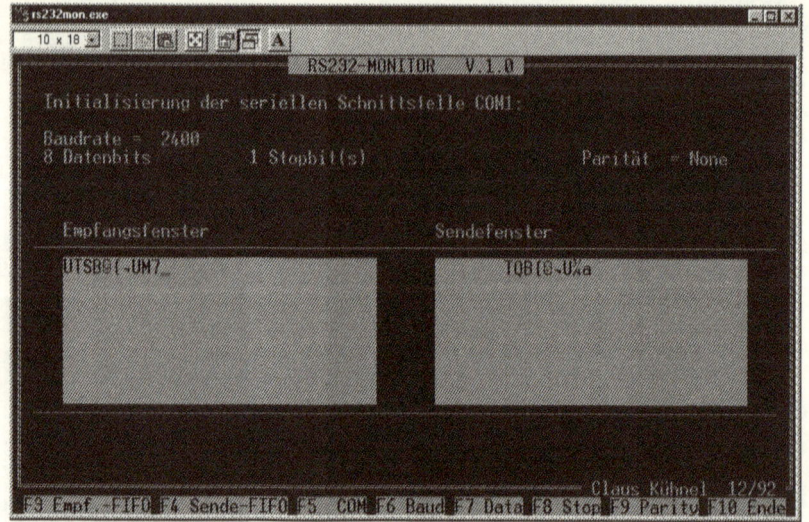

Figure 97 Wrong CRC Byte in S.N.A.P. Telegram

The CRC check in the network node detects the error and sets the acknowledge bits to 11_B. As Figure 98 shows, byte HDB2 in the response is 83 (= &H53) as a result.

SYNC	HDB2	HDB1	DAD	SAD	DB2	DB1	CRC2	CRC1
84	**83**	66	1	123	170	85	**77**	**55**
&H54	**&H53**	&H42	&H01	&H7B	&HAA	&H55	**&H4D**	**&H37**

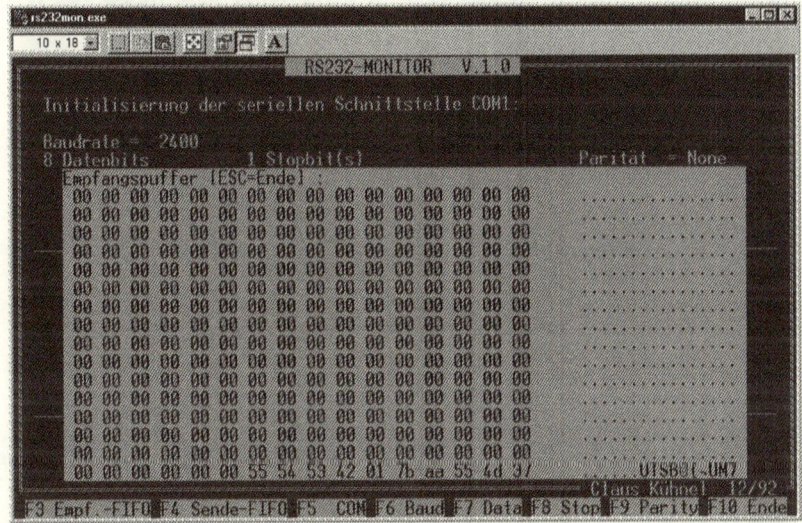

Figure 98 Hexdump of a Received S.N.A.P. Telegram

4.12 CANDIP - Interface to CAN

The German company Bosch developed the "Controller Area Network" (CAN) for the networking of system components in cars. CAN is based on an international standard (ISO 11898). Several semiconductor manufacturers offer CAN controllers and CAN bus drivers.

CAN connects devices featuring equal rights (control devices, sensors, and actors) with a serial bus. In the simplest case, this bus is made up of two wires.

In CAN data transmissions, an identifier known in the whole network characterizes the contents of a message (revolutions or temperature of an engine, for example). There is no addressing of any network node. Besides the characterization of the message contents the identifier determines the priority of the message. The priority is responsible for bus allocation, which is important when several nodes will access the bus.

If a message is to be sent from the CPU of any network node to one or several network nodes, then the data to be sent and the associated identifiers are transferred, together with a request for transmission, to the connected CAN controller. This done, the CPU part is finished.

The generation and transmission of the resulting message is the task of the CAN controller. When the CAN controller gets access to the bus, all other network nodes are receivers of this message.

As soon as the message is received, an acceptance check is performed: the identifier is read, and it is determined whether the data are relevant to this node or not. If so they will be processed, if not they will be ignored.

The contents-related addressing guarantees a high system and configuration flexibility. It is very easy to add new network nodes to an existing CAN network.

The CAN protocol supports two formats of message frames which, essentially, differ in the length of the identifier (ID) only.

The identifier length is 11 bits in the standard format and 29 bits in the enhanced format. The whole message frame for CAN data transmission comprises seven fields. Figure 99 shows a CAN standard frame.

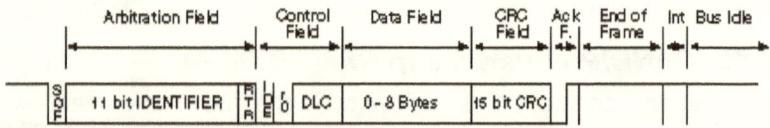

Figure 99 CAN Standard Frame

A standard format message begins with the start bit - Start of Frame (SOF). The Arbitration Field following it contains the identifier (ID) and the Remote Transmission Request bit (RTR). This bit marks the frame as Data Frame or Remote Frame without any data.

The Control Field contains the Identifier Extension bit (IDE) that distinguishes between standard and enhanced format, a reserved bit for further enhancements and the Data Length Code (DLC) specifying the number of data bytes in the frame.

As defined by the DLC, the Data Field can have a length of 0 to 8 bytes.

The CRC Field contains a 15-bit CRC for error detection. The Acknowledge Field (ACK) comprises the ACK slot (one recessive bit). The bit in that ACK slot is sent recessive and will be overwritten dominant from all nodes that receive the message correctly (positive acknowledge). This acknowledge is independent of the result of the acceptance check.

The End of Frame marks the end of a message. Intermission is the minimum number of bit times between two consecutive messages. If there is no further bus access the bus will be idle.

These basics should explain the context and can be consolidated in the relevant CAN literature.

Based on the AT90S8515, the Swedish company LAWICEL [http://www.candip.com] developed the microcontroller module CANDIP/AVR. CANDIP/AVR contains all components required to build an interface to the CAN bus. Figure 100 shows the CANDIP/AVR module.

Figure 100   CANDIP/AVR Microcontroller Module

The CANDIP/AVR microcontroller module has the following features:

- Standard 28 pin DIP board with 0.1" pins (use a standard DIP28 carrier).
- Needs a 5V DC/30mA power source only.
- Atmel AVR type AT90S8515 normally working at 3.6864MHz.
- 8k user FLASH, 512 bytes user RAM and 512 bytes user EEPROM.

- Up to 13 digital I/O points on DIP28 board, each capable of sinking 20 mA as output.
- SPI port for expansion.
- One interrupt line available for user functions (INT1).
- SJA1000 CAN controller working at 16MHz, supporting CAN2.0B.
- 82C250 High Speed CAN transceiver 1Mbit (ISO-11898).
- CAN controller can be interrupt driven (INT0).
- MAX202 RS-232 transceiver which together with the AVR can send/receive up to 115 kbit/s.
- MAX825M reset circuit, the normal RESET and inverted RESET is via external pins.
- No interpreted software, it is programmed with compilers.
- Possibilities to implement higher level protocols such as CANopen, DeviceNet etc.

Figure 101 shows the block diagram of the CANDIP/AVR module with external components.

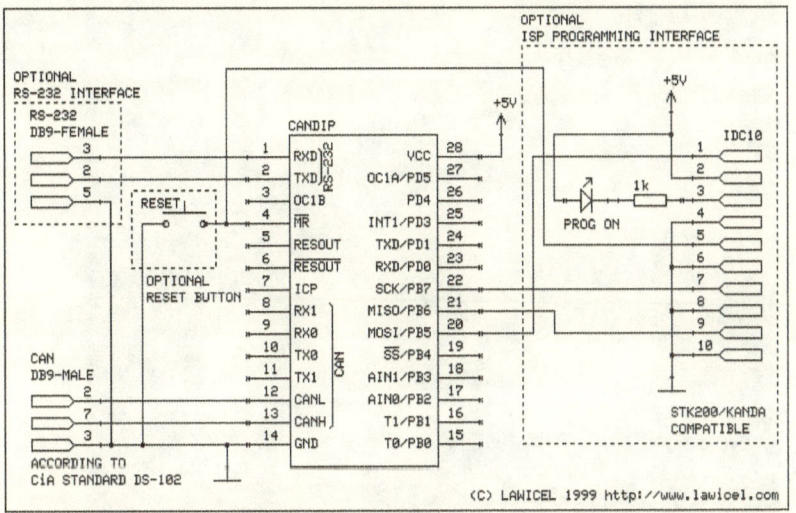

Figure 101
Block Diagram of CANDIP/AVR Module with External Components

Optional components can be used in dependence on the planned application. The CANDIP/AVR module supports In-System-Programming (ISP) via the SPI Port, and the STK200 Programmer from Atmel/Kanda.

For fast and comfortable debugging of CAN applications LAWICEL offers the Activity Board (ACB1) for CANDIP/AVR. Figure 102 shows this Activity Board for CANDIP/AVR.

Figure 102   CANDIP/AVR Activity Board

On the basis of the introduced hardware, a first CAN application can now be developed.

With a minimum of two Activity Boards a CAN network can be created by connecting the two CAN bus lines CAN_Hi and CAN_Lo.

Figure 103 shows the circuitry of our sample network with external components of both CANDIP/AVR network nodes.

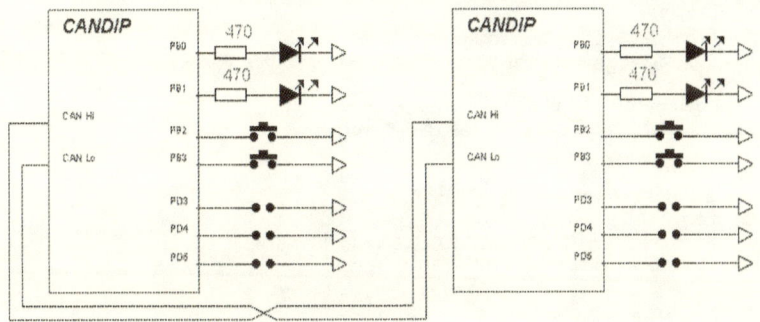

Figure 103 CANDIP/AVR Network

The node identification (NodeId) is set by jumpers at PinD3 to PinD5 and is queried after program start.

PortB serves for I/O. Two keys are connected to PinB2 and PinB3. Pressing any key changes the bit pattern at PinB2 and PinB3 and the program generates a CAN message to inform the network about the new state of inputs PinB2 and PinB3.

The other network node receives this CAN message and displays the bit pattern by means of the LEDs connected to PinB0 and PinB1.

Listing 49 shows the program for each node of our sample network. The three subroutines Initsja, Transmitcanio and Checkcan are important to the CAN bus management.

```
' CAN example by BASCOM-AVR

' Purpose:     General Test routines
'              for SJA1000 on the CANDIP/AVR in BasicCAN mode
'
' Chip:        AT90S8515 running at 3.6864MHz
'
' Version:     1.0.0, 25:th of February 2000
'
' Author:      Lars Wictorsson
'              LAWICEL / SWEDEN
'              http://www.lawicel.com   lars@lawicel.com
'
' Remarks:
' This code sample is provided as is and demonstrates
' simple distributed I/O by CAN.
' The CANDIP is reading two push buttons and sends their
' current status as CAN frames when they are changed.
' The NodeId used is read from the CANDIP Activity board
```

```
' jumpers (PD3-PD5) when started.
' When button PB2 and/or PB3 is pushed/released their
' status is sent on the CANbus based on the NodeId read
' from startup.
' The other node is "listening" for this ID and will
' display the status on the LED's PB0 and PB1 and vice
' versa.
' This demonstrates the Multi Master functionality of CAN.
' This program is tested with BASCOM-AVR version 1.0.0.8.
'
' Test Setup:
' Use 2 CANDIP/AVR's and 2 Activity boards.
' On one Activity board, set PD3-PD5 open (NodeID=0).
' On the other, set PD3 closed, PD4-PD5 open (NodeID=1).
' Set PB0-PB1 as output and PB2-PB3 as input.
'
' Important:
' The MakeInt function in BASCOM is wrong in version
' 1.0.0.8 and will be fixed later, this means you need to
' swap the msb and lsb (the help file in BASCOM shows it
' correct but compiler is wrong, this is a known bug of
' BASCOM).
'
' History:    2000-02-25  1.0.0   Created
'
' CANDIP:     See CANDIP at http://www.lawicel.com/CANDIP
'

$crystal = 3686400
$baud = 57600

' SJA1000 CAN contoller is located at &H4000
Const Can_base = &H4000

' Some SJA1000 registers in BasicCAN mode
Const Can_ctrl = &H4000
Const Can_cmd = &H4001
Const Can_status = &H4002
Const Can_int = &H4003
Const Can_ac = &H4004
Const Can_am = &H4005
Const Can_tmg_0 = &H4006
Const Can_tmg_1 = &H4007
Const Can_ocr = &H4008
Const Can_test = &H4009
Const Can_tx_id = &H400A
Const Can_tx_len = &H400B
Const Can_tx_buf0 = &H400C
Const Can_tx_buf1 = &H400D
Const Can_tx_buf2 = &H400E
Const Can_tx_buf3 = &H400F
Const Can_tx_buf4 = &H4010
Const Can_tx_buf5 = &H4011
Const Can_tx_buf6 = &H4012
Const Can_tx_buf7 = &H4013
Const Can_rx_id = &H4014
Const Can_rx_len = &H4015
Const Can_rx_buf0 = &H4016
Const Can_rx_buf1 = &H4017
Const Can_rx_buf2 = &H4018
Const Can_rx_buf3 = &H4019
```

```
Const Can_rx_buf4 = &H401A
Const Can_rx_buf5 = &H401B
Const Can_rx_buf6 = &H401C
Const Can_rx_buf7 = &H401D
Const Can_clkdiv = &H401F

' Some key values

Const Own_id = 0
' Our CAN-ID
Const Acceptmask = &HFF
' Our accept mask

' Some useful bitmasks

Const Resreq = 1
' Reset Request
Const Rbs = 1
' Receive Buffer Status
Const Rrb = 4
' Release Receive Buffer
Const Txreq = 1
' Transmit Request
Const Tba = 4
' Transmit Buffer Access

Declare Sub Initsja
Declare Sub Transmitcanio( b as byte)
Declare Sub Checkcan

Dim Always As Byte
Dim Nodeid As Byte
Dim Inpb As Byte
Dim Inpbold As Byte

Always = 1
Inpb = &H0C
' Default button status
Inpbold = &H0C

Mcucr = &HC0
' Enable External Memory Access With Wait - state

Ddrb = &H03
' Set PB0+PB1 as output and PB2+PB3 as input with pull-up
Portb = &H0F
' and turn off LED's

Ddrd = &H00
' Set PD3+PD4+PD5 as inputs with pull-up
Portd = &H38

Nodeid = Pind
' Read Jumper inputs on Port D and save as Node ID.
Rotate Nodeid , Right , 3
Nodeid = Nodeid And &H07
Nodeid = 7 - Nodeid
' Invert, how to make it better in BASCOM?
```

```
Initsja

While Always = 1
  Inpb = Pinb And &H0C
' Read inputs PB2 & PB3

  If Inpb <> Inpbold Then
' Are they different from last check?
    Transmitcanio Inpb
' If so, send new state of buttons
    Inpbold = Inpb
' and save this state
  End If
  Checkcan
Wend

End

Sub Initsja
' Initiate CAN controller 125kbit
  Local B As Byte
  B = Inp(can_ctrl)
  B = B And Resreq
  While B = 0
    out can_ctrl,resreq
    B = Inp(can_ctrl)
    B = B And Resreq
  Wend
  out Can_ac, Own_id
  out Can_am, Acceptmask
  out Can_tmg_0,3
  out Can_tmg_1,&H1C
  out Can_ocr,&HDE
  out Can_clkdiv,7
  out Can_ctrl,&H5E
  out Can_cmd,&H0C
End Sub

Sub Transmitcanio( b as byte)
  Local Id As Word
  Local Tmp1 As Word
  Local Ln As Byte
  Local Tmp2 As Byte

  Do
' Loop until transmit buffer is empty
    Tmp1 = Inp(can_status)
    Tmp1 = Tmp1 And Tba
  Loop Until Tmp1 = Tba

  Id = &H500 + Nodeid
' Create ID based on NodeId
  Ln = 1
  Tmp1 = Id
  Rotate Tmp1 , Right , 3
  Tmp2 = Low(tmp1)
  out Can_tx_id, Tmp2
  Tmp1 = Id And &H07
  Rotate Tmp1 , Left , 5
  Tmp1 = Tmp1 + Ln
  Tmp2 = Low(tmp1)
```

```
   out Can_tx_len, Tmp2
   out Can_tx_buf0, b
   out Can_cmd, Txreq
End Sub


Sub Checkcan
  Local Id As Word
  Local Tmp1 As Word
  Local Ln As Byte
  Local Tmp2 As Byte

  Tmp2 = Inp(can_status)
  Tmp2 = Tmp2 And Rbs

  If Tmp2 = Rbs Then
    Tmp2 = Inp(can_rx_id)
    Id = Makeint(0 , Tmp2)
    Rotate Id , Left , 3
    Tmp1 = Inp(can_rx_len)
    Rotate Tmp1 , Right , 5
    Tmp1 = Tmp1 And &H07
    Id = Id + Tmp1
    Tmp2 = Inp(can_rx_len)
    Ln = Tmp2 And &H0F
    Tmp2 = Inp(can_rx_buf0)
    Rotate Tmp2 , Right , 2
    If Nodeid = 0 Then
      If Id = &H501 Then
        Portb = &H0C + Tmp2
      End If
    Elseif Nodeid = 1 Then
      If Id = &H500 Then
        Portb = &H0C + Tmp2
      End If
    End If
    out can_cmd, rrb
' Release receive buffer
  End If
End Sub
```

Listing 49   CAN Test Program (CANDIPIO.BAS)

Before initializing the CAN controller SJA1000 (Philips) it must be put to the Reset Mode. Thereafter, the initial values can be written to the Control Segment. The data transfer rate is here set to 125 kbit/sec.

By evaluating the identifier of the CAN messages received, the Acceptance Filter decides which CAN messages will be saved in the receive buffer (RXFIFO). In the initialization, the Acceptance Filter is transparent. All received CAN messages are saved in RXFIFO.

To change the initialization, it is absolutely necessary to consult data sheet "SJA1000 Stand-alone CAN Controller" [http://www.semiconductors.philips.com].

Subroutine `Transmitcanio` sends the CAN message to the network. After the subroutine call, the routine waits until the Transmit Buffer Status signalizes a free buffer. When the Transmit Buffer is free, the CPU can write a prepared message to the buffer.

Preparing the CAN message means defining identifier, data length, and data bytes. According to Figure 103 the identifiers are &H500 and &H501. The data length is one byte for the input state of the two input lines.

These definitions are followed by the output of identifier, data length, data byte, and a Transmission Request. The Transmission Request requests the CAN controller to send this CAN message.

All received CAN messages that have passed the Acceptance Filter are written to the Receive Buffer. Subroutine `Checkcan` checks the Receive Buffer for CAN messages and processes them, if necessary. The subroutine reads identifier, data length, and data byte from the Receive Buffer.

If the received CAN messages came from the respectively other node, the transmitted input state is displayed by the connected LEDs. After the received CAN message has been processed, the Receive Buffer is released again.

On the basis of the described program example CANDIPIO.BAS, further CAN applications can be developed using BASCOM-AVR.

A lot of supporting hardware is now available if an 8051 derivative is preferred to be used for the CAN application. There are 8051 derivatives with integrated CAN controllers or modules comparable with CANDIP/AVR.

Based on Infineon's C505CA, LAWICEL is offering the CANDIP/505. Features of the CANDIP/505 microcontroller module:

- Standard 28 pin DIP board with 0.1" pins (use a standard DIP28 carrier).
- 4 layer board for good EMI performance.
- Needs a 5V DC/30mA power source only (plus 70mA for CAN transceiver).
- Infineon type C505CA working at 16MHz.
- 64k bytes user FLASH, 1k bytes user XRAM and 128 bytes user EEPROM.
- ADC with 10bit resolution / 4 channels.

- Software controller SPI port for expansion.
- On chip full CAN-Controller (CAN 2.0B).
- 82C250 High Speed CAN transceiver 1Mbit (ISO-11898).
- MAX202 RS-232 transceiver.
- MAX825M Reset circuit.
- No interpreted software, it is programmed with compilers.
- Possibilities to implement higher level protocols such as CANopen, DeviceNet, etc.
- PC-Bootloader for program download in Flash-EPROM via the CAN interface.
- Demo software for the individual hardware components.

For fast and comfortable debugging of CAN applications, LAWICEL offers the Activity Board (ACB2) for CANDIP/505.

4.13 Random Numbers

Random numbers are numbers which are not created as a result of a functional context (formula or function, for example); their values are purely coincidental.

From a physical point of view noise sources generate a signal more or less randomly. After digital-to-analog conversion such a signal can serve as a random number.

In most cases pseudo-random number generators are used. They operate according to defined rules but the results are random.

Program RANDOM.BAS is a simple pseudo-random number generator later used for test purposes. Listing 50 shows the source of program RANDOM.BAS.

```
Dim Value As Integer
Dim Seed As Integer

Declare Function Random(byval Z As Integer) As Integer

Seed = 1234                       ' or other initialization value
                                  ' from RTC for example
Value = Random(1000)              ' calculates a pseudo random
                                  ' numer between 0 and 1000
End


Function Random(byval Z As Integer) As Integer
   Local X As Integer
   Local Y As Long

   X = Seed * 259
   X = X + 3
   Seed = X And &H7FFF
   Y = Seed * Z
   Y = Y / &H7FFF
   Y = Y + 1
   Random = Y
End Function
```

Listing 50   Generation of random numbers (RANDOM.BAS)

Function `random()` is the core of program RANDOM.BAS. The parameter of that function defines the range of the random number to be generated. Function call `Value = Random(1000)` will generate a random number between 0 and 1000.

If the program is restarted, the same sequence of random numbers will be generated. For test purposes this is a preferred feature, but not so for applications.

In playing dice it would not be very thrilling if the numbers could be predicted.

Variable `Seed` defines the random number sequence and must be initialized before the first call of function `random()`. Normally, `Seed` is initialized with the same value after each program start, and the same sequence of numbers will be generated.

If `Seed` is initialized with a random number, another sequence of random numbers will be generated after each program start. Using a connected real-time clock for initialization of variable `Seed` is one solution to avoid that always the same sequence of numbers occurs.

For test purposes random numbers can be sent via a serial port to the microcontroller. A terminal program can send a data file containing random numbers to the microcontroller that can use these number as measuring results, for example.

A good source of random numbers is the URL http://www.random.org/ nform.html. The number of random numbers as well as the minimum and maximum values can be defined in an input form. Figure 104 shows such a (completed) input form.

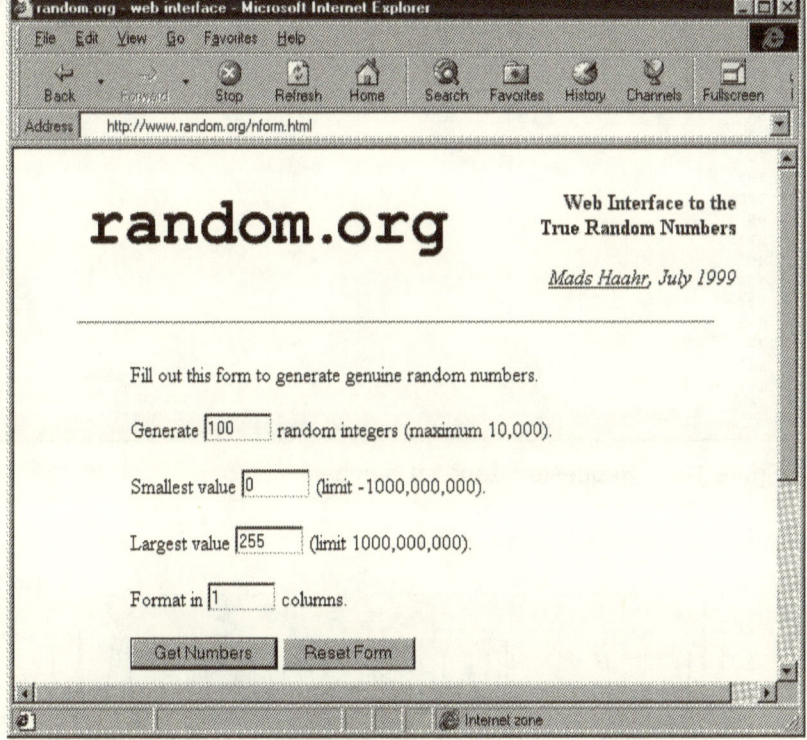

Figure 104   Input Form for Random Number Generation

In the example, 100 random numbers between 0 and 255 shall be generated in one column. Figure 105 shows the result of the requested operation. Figure 106 shows the generated random numbers in a graphic presentation.

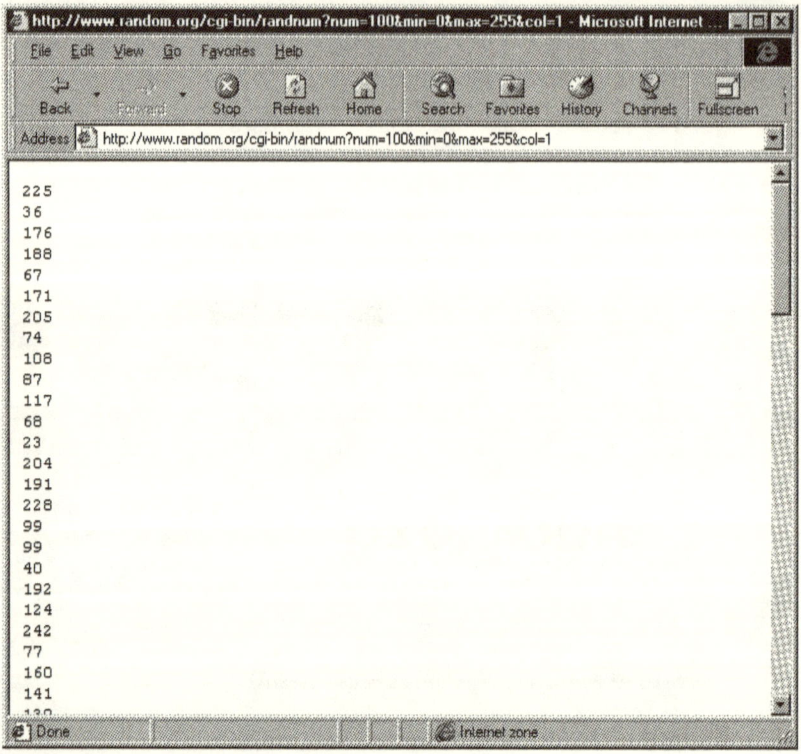

```
225
36
176
188
67
171
205
74
108
87
117
68
23
204
191
228
99
99
40
192
124
242
77
160
141
```

Figure 105 Requested Random Numbers

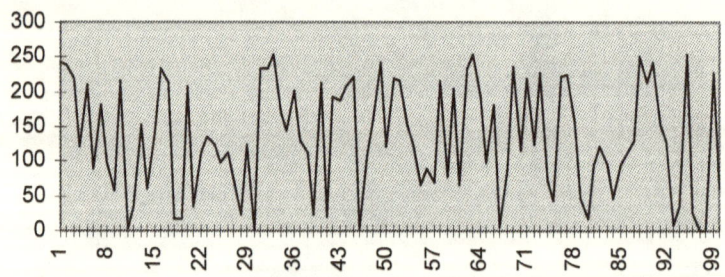

Figure 106 Graphic Presentation of Requested Random Numbers

To save the generated random numbers for later use, the contents of the browser window are saved as text file. The proposal in the browser was RANDNUM.TXT and is used without any changes.

Important is the saving as text file, because the alternative HTML file contains a lot of information not used here.

A terminal program can send this text file to the microcontroller now. A simple program was installed for test purposes (Listing 51). The program will be ported to BASCOM-8051 when PortB is changed to P3, for example, and line `Config Portb` ... is erased.

```
' GETRANDOM.BAS by BASCOM-AVR

Dim Value As Byte
Dim Str_input As String * 4

Config Portb = Output
Portb = &HFF

Do
    Input Str_input Noecho
    Value = Val(str_input)
    Portb = Not Value
Loop

End
```

Listing 51   Test Program GETRANDOM.BAS

Input string `Str_input` is read from the serial port in an endless loop. After conversion to a numeric value, the LEDs connected to PortB will display the respective value. Each random number saved in RANDNUM.TXT has max. three digits and is supplemented by a carriage return CR (&H0D).

Due to parameter `Noecho` in instruction `Input Str_input Noecho`, GETRANDOM.BAS sends no characters back to the terminal.

Normally, other things are done with random numbers. In the program example, the generation and provision of random numbers was important first and foremost. The next paragraph will demonstrate the use of random numbers for the test of an algorithm.

4.14 Moving Average

The calculation of the moving average is a basic function of many measuring instruments.

Each measured value has a variation range. In an audio signal such a variation can be heard as noise. The calculation of a moving average is one means to reduce, or suppress, noise in any signal.

Figure 107 shows the principle of calculating a moving average.

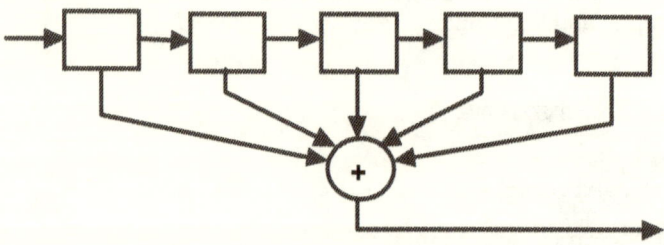

Figure 107 Calculating the Moving Average

Sampled data (measured values) will be shifted through a shift register. In Figure 107 the actual sample is saved in the extreme left position of the shift register after all saved data have been shifted one position to the right. The extreme right position of this shift register – it contains the oldest value – is overwritten in every shift operation.

A shift register of length n can save the last n sampled data. These last n samples can be used for the calculation of an average (mean value). To calculate the average, add up all values and divide by n. This procedure is to be repeated with every new sample.

In the next program example MEAN.BAS. a sampled measured value, is replaced by a random number. In this way, the simulator can be used to test the whole procedure. Listing 52 shows the source of this program example.

```
' MEAN.BAS by BASCOM-AVR

Dim Mean As Byte
Dim Mean_temp As Integer
Dim Lenght As Word
Dim Index As Byte
Dim Temp As Byte

Dim Value As Integer
Dim Seed As Integer

Declare Function Random(byval Z As Integer) As Integer

Lenght = 5
Dim Buffer(lenght) As Byte

Config Portb = Output
Portb = &HFF

Seed = 1234                        ' or other initialization value
                                   ' from RTC for example

For Index = 1 To Lenght
  Buffer(index) = 0
Next

Do
  Index = Lenght - 1
  Do
    Temp = Buffer(index)
    Incr Index
    Buffer(index) = Temp
    Decr Index
    Decr Index
  Loop Until Index = 0

  Value = Random(&Hff)             ' calculates a pseudo random
  Buffer(1) = Low(value)           ' number between 0 and 255
                                   ' and write it to buffer
  Mean_temp = 0                    ' calculate mean value
  For Index = 1 To Lenght
    Mean_temp = Mean_temp + Buffer(index)
  Next
  Mean_temp = Mean_temp \ Lenght
  Mean = Low(mean_temp)

  Portb = Not Mean                 ' display mean value
  Print Value ; " " ; Mean
  Waitms 20
Loop

End
```

```
Function Random(byval Z As Integer) As Integer
  Local X As Integer
  Local Y As Long

  X = Seed * 259
  X = X + 3
  Seed = X And &H7FFF
  Y = Seed * Z
  Y = Y / &H7FFF
  Y = Y + 1
  Random = Y
End Function
```

Listing 52 Moving Average (MEAN.BAS)

The random number generator introduced in program RANDOM.BAS generates the values emulating the sampled data.

Figure 108 shows the sequence of random numbers and the moving average over five positions.

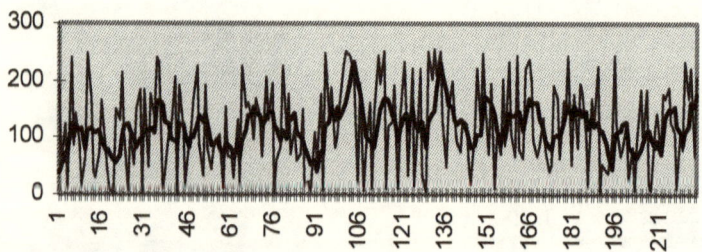

Figure 108 Moving Average over Five Positions

A better smoothing of the measured noise data is obtained with more considered positions. Figure 109 shows the same sequence of random numbers and the moving average over 16 positions.

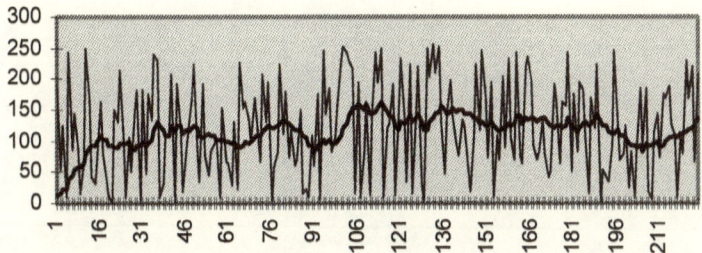

Figure 109 Moving average over 16 positions

Better smoothing gives a more even curve but must be paid for with a delay.

Choosing the length of the shift register as power of 2 (2, 4, 8 etc.) allows the awkward division operation to be replaced by a simple shift operation.

For a shift register length of 16, the division by 16 can be replaced by a shift by 4:

```
'Mean_temp = Mean_temp \ Length
Shift Mean_temp , Right  ,   4
```

Such an adaptation reduces the run time of a program considerably. While the division in the above example takes 72.25 μs, the shift operation will already be finished after 19.75 μs.

5 Appendix

5.1 Decimal-Hex-ASCII Converter

The following Decimal-Hex-ASCII Table supports the conversion of the different data formats.

DEC	HEX	ASCII	Key	DEC	HEX	ASCII	Key
0	0x00	NUL	Ctrl @	64	0x40	@	
1	0x01	SOH	Ctrl A	65	0x41	A	
2	0x02	STX	Ctrl B	66	0x42	B	
3	0x03	ETX	Ctrl C	67	0x43	C	
4	0x04	EOT	Ctrl D	68	0x44	D	
5	0x05	ENQ	Ctrl E	69	0x45	E	
6	0x06	ACK	Ctrl F	70	0x46	F	
7	0x07	BEL	Ctrl G	71	0x47	G	
8	0x08	BS	Ctrl H	72	0x48	H	
9	0x09	HT	Ctrl I	73	0x49	I	
10	0x0A	LF	Ctrl J	74	0x4A	J	
11	0x0B	VT	Ctrl K	75	0x4B	K	
12	0x0C	FF	Ctrl L	76	0x4C	L	
13	0x0D	CR	Ctrl M	77	0x4D	M	
14	0x0E	SO	Ctrl N	78	0x4E	N	
15	0x0F	SI	Ctrl O	79	0x4F	O	
16	0x10	DLE	Ctrl P	80	0x50	P	
17	0x11	DC1	Ctrl Q	81	0x51	Q	
18	0x12	DC2	Ctrl R	82	0x52	R	
19	0x13	DC3	Ctrl S	83	0x53	S	
20	0x14	DC4	Ctrl T	84	0x54	T	
21	0x15	NAK	Ctrl U	85	0x55	U	
22	0x16	SYN	Ctrl V	86	0x56	V	
23	0x17	ETB	Ctrl W	87	0x57	W	
24	0x18	CAN	Ctrl X	88	0x58	X	
25	0x19	EM	Ctrl Y	89	0x59	Y	
26	0x1A	SUB	Ctrl Z	90	0x5A	Z	
27	0x1B	ESC	Ctrl [	91	0x5B	[	
28	0x1C	FS	Ctrl \	92	0x5C	\	
29	0x1D	GS	Ctrl]	93	0x5D	]	
30	0x1E	RS	Ctrl ^	94	0x5E	^	
31	0x1F	US	Ctrl _	95	0x5F	_	
32	0x20	SP		96	0x60	`	
33	0x21	!		97	0x61	a	
34	0x22	"		98	0x62	b	
35	0x23	#		99	0x63	c	

36	0x24	$		100	0x64	d	
37	0x25	%		101	0x65	e	
38	0x26	&		102	0x66	f	
39	0x27	'		103	0x67	g	
40	0x28	(		104	0x68	h	
41	0x29	)		105	0x69	i	
42	0x2A	*		106	0x6A	j	
43	0x2B	+		107	0x6B	k	
44	0x2C	,		108	0x6C	l	
45	0x2D	-		109	0x6D	m	
46	0x2E	.		110	0x6E	n	
47	0x2F	/		111	0x6F	o	
48	0x30	0		112	0x70	p	
49	0x31	1		113	0x71	q	
50	0x32	2		114	0x72	r	
51	0x33	3		115	0x73	s	
52	0x34	4		116	0x74	t	
53	0x35	5		117	0x75	u	
54	0x36	6		118	0x76	v	
55	0x37	7		119	0x77	w	
56	0x38	8		120	0x78	x	
57	0x39	9		121	0x79	y	
58	0x3A	:		122	0x7A	z	
59	0x3B	;		123	0x7B	{	
60	0x3C	<		124	0x7C	|	
61	0x3D	=		125	0x7D	}	
62	0x3E	>		126	0x7E	~	
63	0x3F	?		127	0x7F	DEL	DEL

5.2 DT006 Circuit Diagram

The DT006 board will program the 8, 20, and 28 pin DIP chips on board. There are sockets for all three chips.

You will need a DB-25-male to DB-25 female cable with at least pins 2, 4, 5, 11, and 25 (GND) connected straight through between the DB-25 male and the DB-25 female. Standard DB-25 male to female extension cables that have all 25 wires connected straight through, are fine for this job.

The whole DT006 circuitry is shown on the next page.

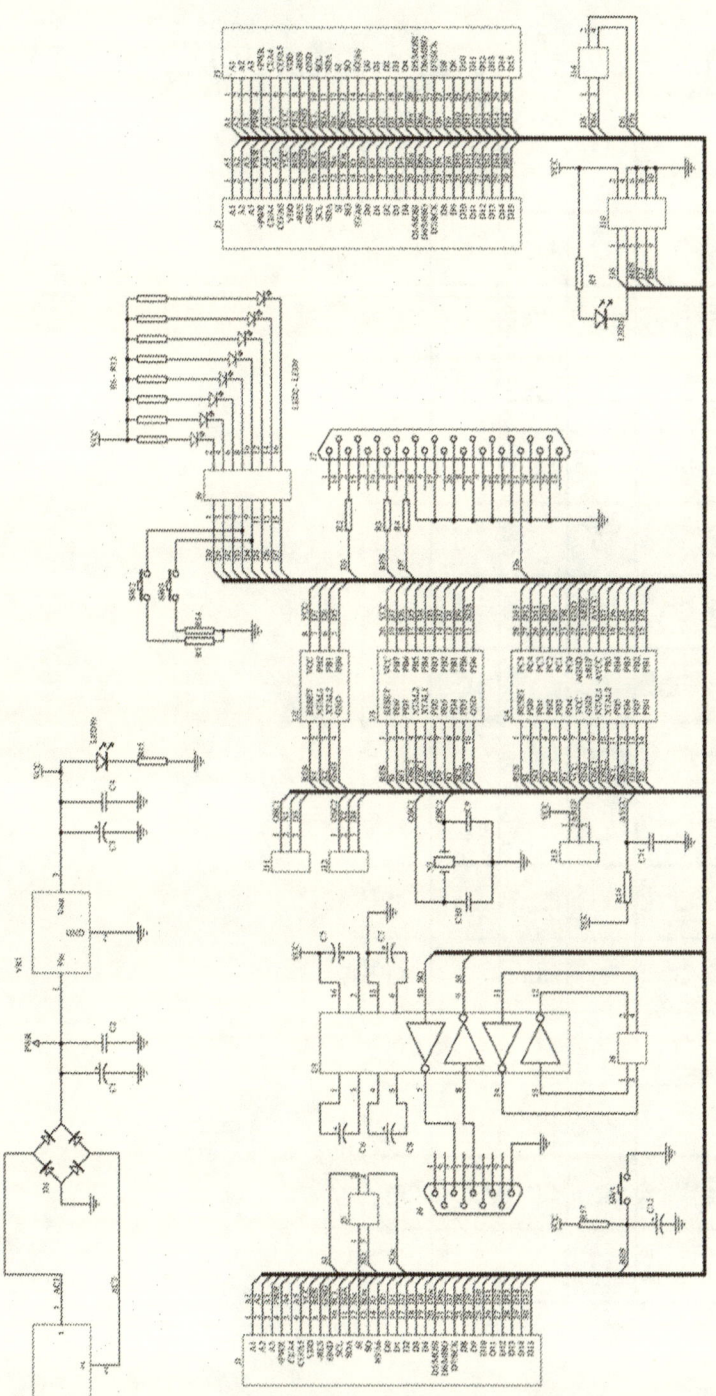

5.3 Characters in Seven-Segment Display

5.4 BASIC Stamp II

The BASIC Stamp II (BS2) is a microcontroller programmable in BASIC. Based on Microchip's PIC16C57 and equipped with Parallax's PBASIC2 Firmware, a microcontroller is obtained that executes BASIC tokens saved in an EEPROM. The whole infrastructure is available in a 24pin DIL module. Figure 110 shows the BS2 Module.

Figure 110   BS2 Module

In addition to the BS2 Chip programmed with PBASIC firmware, the BS2 Module contains an I²C EEPROM, a reset device and a voltage regulator.

Parallax offers a complete PC development environment for download [http://www.parallaxinc.com].

A program prepared for BS2 can be downloaded from PC to BS2 via a serial link. After reset, the program will run on BS2.

See Parallax's or the author's website [http://www.ckuehnel.ch] for further information on the BASIC Stamp.

5.5 Literature

[1] EDN's 25th Annual Microprocessor/Microcontroller Directory
EDN Access, September 24, 1998
http://www.ednmag.com/reg/1998/092498/8_8051.htm

[2] C8051Fxxx Mixed-Signal Microcontroller Family Datasheets
Cygnal Integrated Products, Inc
http://www.cygnal.com/view.asp?page=datasheets

[3] Turley, J.:
Atmel AVR brings RISC to 8-bit World.
Microprocessor Report, Vol. 11, H. 9, Sunnyvale/CA 1997.

[4] Kühnel, C.:
AVR RISC Microcontroller Handbook.
Newnes: Boston, Oxford, Johannesburg, Melbourne, New Delhi, Singapore, 1998

[5] AVR313: Interfacing the PC AT Keyboard.
Atmel Application Note

[6] Kühnel, C.; Zahnert, K.:
BASIC Stamp. 2.Edition
Newnes: Boston, Oxford, Johannesburg, Melbourne, New Delhi, Singapore, 2000

5.6 Links

Author's Web Site - Distribution of BASCOM in D, CH and A
http://www.ckuehnel.ch
MCS Electronics' Website - Developer of BASCOM
http://www.mcselec.com
BASCOM Forum
http://ch.onelist.com/community/BASCOM

Equinox Technologies
Programmers and Evaluation Modules for 8051 and AVR
http://www.equinox-tech.com

Dontronics
The World's Largest Range of Atmel/AVR & PICmicro HW and SW
Free Basic Compiler and Programmer
http://www.dontronics.com/runavr.html
The Little "rAVeR!" AVR & Basic Kit
http://www.dontronics.com/dt006.html

Practical Tips on Serial Communication
http://www.seetron.com/ser_an1.htm
DS1820 1–Wire™ Digital Thermometer
DS1820 Data Sheet
http://www.dalsemi.com/DocControl/PDFs/1820.pdf
MicroLAN Design Guide
Description of 1-Wire Networks from Dallas Semiconductors
http://www.dalsemi.com/TechBriefs/tb1.html
Understanding and Using Cyclic Redundancy Checks with Dallas Semiconductor iButton™ Products
Description of CRC Checks for 1-Wire Components
http://www.dalsemi.com/DocControl/PDFs/app27.pdf

DECODING IR REMOTE CONTROLS
Decoding the RC5 Commands includes a sample program for 8052 in assembler.
http://www.ee.washington.edu/eeca/text/ir_decode.txt
The RC5 code, Philips
Description of RC5 Commands
http://kwik.ele.tue.nl/pvdb/rc/philips.html

S.N.A.P - Scaleable Node Address Protocol
Description of S.N.A.P. including some sample programs and possibility of download
http://www.hth.com/snap/
CANDIP - How easy and inexpensive can CAN be?
LAWICEL's Web Site with description of the CANDIP device
http://www.candip.com
SJA1000 Stand-alone CAN Controller
Philips' web site for download of CAN controller data sheet
http://www.semiconductors.philips.com

6 Index

$EXTERNAL 47
$LIB 47
16-Bit Timer/Counter 81
1-Wire 151
1-Wire Bus System 151
1-Wire Network 151
1WIRE1.BAS AVR 156
1WIRE2.BAS AVR 158
1WIRE3.BAS 160
3x4 Keypad 133
4-Bit Mode 32
5 x 7 LED matrix 114
7SEGMENT.BAS 8051 112
8051 Assembler 75
8-Bit Mode 32
Acceptance Filter 207
Acknowledge 171
Alphanumeric LCD 42
Analog Comparator 16
Applications 77
Architecture 12
Associated Function 180
Asynchronous serial communication 16; 143
AT90S8515 13
AT90S8515 as a logical device 78
ATKBD.BAS AVR 139
AutoUpdate 25
AVR Assembler 74

AVR ICP910 Programmer 50
AVR ISP Programmer 36
AVR microcontroller 11
AVR Studio 30; 64
BASCOM demos 23
BASCOM-AVR Demo Files 24
BASCOM-AVR Options 28
BASIC Compiler 23
BASIC Stamp 148; 223
Block diagram 8051 10
Block diagram AT90S8515 14
Block diagram C8051F0000 11
Block diagram of 8051 timer .. 105
Block Diagram of CANDIP/AVR Module 201
BlockMove Routine 47
BS2 .... 148; 223
BS2 Module 223
Building new instructions 69
Bus Master 151
BYREF 71
Byte Write Operation 171
BYVAL 71
Calculation of 8-bit CRC 160
CAN 197
CAN Controller SJA1000 207
CAN Standard Frame 198
CANDIP Activity Board .. 201; 209

227

CANDIP/AVR Microcontroller Module 199	Debugging 9; 27; 60; 209
CANDIP/AVR Network........... 203	Decimal-Hex-ASCII-Converter 219
CANDIPIO.BAS AVR............. 207	Device Type Identifier 169
CAPTURE1.BAS 90	Differences between AVR and 8051 59
Capturing a pulse length 103	Digital Thermometer 152
Character Generator RAM..... 120	Digital-to-analog conversion 94
Checkcan 203; 208	Digital-to-Analog Conversion by PWM 99
Chip>Autoprogram 66	Dot-matrix Display 114
CK.LIB 48	DOTMATRIX.BAS AVR 118
Clock Generator 82	Download 23; 144
Clock Select Bits 92	DPTR 9
Commercial version........... 23; 25	DS1820 152
Compare Interrupt 84	DS1820 Block Diagram 152
Compiler Directives 69	Duty 92
Configuration of On-Chip SPI 164	Error messages 25
Connecting LEDs 108	Evaluation board MCU00100 ... 50
Connection of a Matrix Keypad 134	Extended Keys 137
Connectors for PC-AT keyboard 137	Family Code 152
Controller Area Network 197	FBPRG Programmer 51
Controlling an LCD 32	*File>New* 27
Counter Mode 89	*File>Open* 27; 64
COUNTER0.BAS 89	Frame 28
CPHA 166	Getatkbd() 137
CPOL 166	GETKBD() 135
CRC16 Error Detection 175	GETRANDOM.BAS 213
CRC-Byte 152	GETRC5() 142
Customer-specific characters .. 42	Global Interrupt Enable/Disable 15
Data Direction Register 16	Graphic BMP Converter........... 37
Data Display RAM 120	Harvard architecture 12
Data-pointer 9	HD 44780 42
Debug>Go 65	Header Definition Bytes . 175; 176

Hello World.............................57	LCD.BAS 38; 122
Help System 67	LCD1.BAS 44
I/O Ports 16	LCDs with a serial interface...122
I²C Bus Network 168	LED..107
I²C-Bus 167	Library....................................46
I²C-Bus EEPROM..................169	Library Manager................37; 46
ICP910 Programmer 66	Link to external programmer....53
IIC.BAS AVR 173	Local variable 28
Initsja....................................203	LOGIC.BAS 8252 81
Inkey() 146	LOGIC.BAS AVR 79
Input Capture..........................16	LOGIC1.BAS AVR 80
Input Capture Noise Canceler .91	Logical devices 77
Input Form for Random Number Generation.........................211	Low Pass 98
Installation 23	Manual 23
Instruction set........................17	Matrix Keypad 132
Internal resources..................13	MAX232................................143
Interrupt module 15	MCS.LIB 46
Interrupt vector 15	MEAN.BAS 216
Interrupt vector table 15	Memory maps AT90S8515 15
IR Receiver SFH506-36 142	Microcontroller Project.............9
IR remote control..................140	Micro-Pro 51 52
KEY1.BAS 8051 132	MISO....................................161
KEY1.BAS AVR 131	Mixing of BASIC and Assembler73
KEY2.BAS AVR 135	MOSI....................................161
Keyboards 128	Moving Average....................214
Keypad 1x12........................129	Multi-Master System 167
Keypads 128	Networking...........................173
LCD Controller HD44780119; 123	NM25C04..............................162
LCD Designer............37; 42; 115	On-Chip RAM 9
LCD in bus mode..................120	***Options>***
LCD SetUp 33	***Compiler>Communication*** 30

Options>Communication 33	Pulse-Width-Modulation 16
Options>Communications 41	PULSIN.BAS 101
Options>Compiler>Chip 28	PULSIN1.BAS 103
Options>Compiler>I2C, SPI, 1WIRE 31; 164	PULSIN2.BAS 103
Options>Compiler>LCD 32	PWM 16; 92
Options>Compiler>Output ... 29	PWM0.BAS 94
Options>Environment 34	PWM1.BAS 95
Options>Programmer 36	Random Numbers 209
Options>Programmer>Other 52	Random Read Operation 171
Options>Simulation Options 65	RANDOM.BAS 210
Options>Simulator 35	RC5 140
Output Compare 16	RC5 Device Address 141
Output Compare Function 85	RC5.BAS AVR 142
Output Compare Mode 82	Report file 63
Parameter handling 71	Reportfile 64
Parameter passing BYREF 73	RS-232 Level Conversion 144
Parameter passing BYVAL 72	Runtime measurement 49
PC-AT Keyboard Interface 138	RXFIFO 207
PC-AT keyboards 136	S.N.A.P. 174
Peripherals 3; 9; 11	S.N.A.P. Monitor 179
Pin configuration AT90S8515 .. 17	SC 167
Prescaler 16	Scalable Network 174
Program RS232MON 193	Scan Codes of a PC-AT Keyboard 137
Program SnapLab 190	Scratchpad RAM 9; 153; 158
Program>Compile 62	SDA 167
Program>Send to Chip 66	Segment Control 110
Program>Show Result 62	Serial Peripheral Interface 16
Program>Simulate 38; 64	SERIAL.BAS 41
Program>Syntax Check 61	SERIAL.BS2 150
Programming mode 17	SERIAL1.BAS 144
Programming with AVR ICP91050	SERIAL2.BAS 145
	SERIAL3.BAS 146

SERIAL4.BAS AVR 149	SW_UART.BAS AVR 126
Seven-Segment Display in BASCOM-8051 Simulator .. 113	Telegram 175
Seven-segment displays 108	Terminal Emulator 37; 40
SIM_TIMER.BAS 59	Terminal Emulators 34
SIM_TIMER.BAS for 8051 59	TEST_LIB.BAS 49
SIM_TIMER.BAS for AVR 58	Timer Period 57; 58; 82
SIM_TIMER.RPT for 8051 64	TIMER.BAS 8051 107
SIM_TIMER.RPT for AVR 63	Timer/Counter 16; 81
Slave 151	Timer0 82
Smoothing 216	TIMER0.BAS 85
SNAP-IO.BAS 190	TIMER0_1.BAS 87
SNAP-MON.BAS 183	Timer1 82
Software UART 147	TIMER3.BAS 84
SPI Clock Rate 166	*Tools>LCD Designer* 42
SPI Control Register 165	*Tools>Terminal emulator* 41
SPI Interface 161	Transmitcanio 203
SPI Timing 162	UART 16; 143
SPI.BAS AVR 163	Up-counter 84
SPI1.BAS AVR 164	Update of BASCOM-AVR 26
SPI4.BAS AVR 165	*View>New Memory View* 65
Stack size 28	*View>Peripheral>Port>PortB* 65
Subroutine Construct 69	Waitkey() 146
SW_UART.BAS 8051 128	Watchdog Timer 16
	X051 Demo Module 54

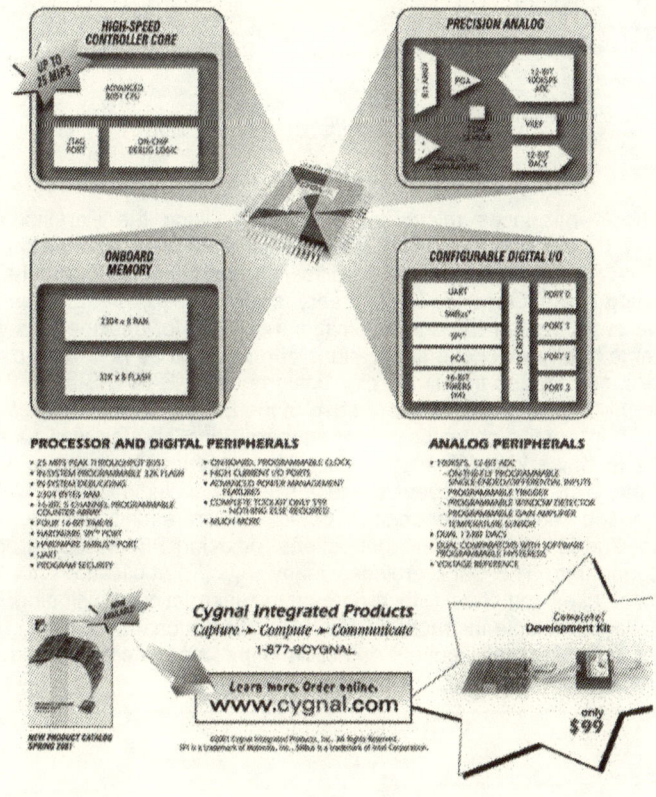

Other books by Claus Kuhnel:

Introduction to PIC microcontroller operation. Applications for designers and hobbyists. Integrated hardware and software coverage.

ISBN: 0-7506-7245-5

Softcover
Measurements: 6 x 9.25 In.
Pages: 305 pp

Publication Date: July 21, 2000

Price: $29.95

This book introduces microcontroller theory using the Parallax BASIC Stamp I, II, and IIsx. The BASIC Stamp microcontroller is based on Microchip's PIC hardware with some modifications and is very approachable for beginning users. Once the basic theory is established, BASIC Stamp, 2nd Ed. walks the reader through applications suitable for designers as well as the home hobbyist. These applications can be used as is or as a basis for further modifications to suit specific design needs. BASIC Stamp, 2nd Ed. thoroughly explains the hardware base of the BASIC Stamp microcontroller including internal architecture, the peripheral functions, as well as providing the technical data sheets for each kind of chip. The authors also explain the BASIC Stamp development systems including DOS and Windows-based tools in tremendous detail. As an added feature, BASIC Stamp, 2nd Ed. includes full instructions for using PBASIC programming and formatting. The book provides many specific applications for microcontroller use, complete with programming instructions, including: single instructions, multiple instructions, interfacing directions, and more complex applications such as motion detection, light measurement, and home automation.

Practical guide for advanced hobbyists or design professionals.
Development tools and code available on the Web.

ISBN: 0-7506-9963-9

Paperback
Measurements: 6 x 9.25 In.
Pages: 256pp

Publication Date: May 07, 1998

Price: $36.95

The AVR RISC Microcontroller Handbook is a comprehensive guide to designing with Atmel's new controller family, which is designed to offer high speed and low power consumption at a lower cost. The main text is divided into three sections: hardware, which covers all internal peripherals; software, which covers programming and the instruction set; and tools, which explains using Atmel's Assembler and Simulator (available on the Web) as well as IAR's C compiler.

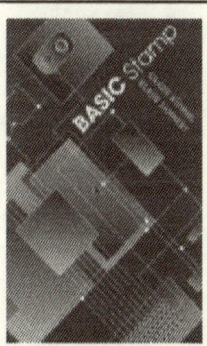

Introduction to PIC microcontroller operation.
Applications for designers and hobbyists.
Integrated hardware and software coverage.

ISBN: 0-7506-9891-8

Paperback
Measurements: 6.25 x 9.25 In.
Pages: 208pp

Publication Date: February 19, 1997

Price: $34.95

BASIC Stamp introduces microcontroller theory using the Parallax BASIC Stamp I and II. The BASIC Stamp microcontroller is based on Microchip's PIC hardware with some modifications, and is very approachable for beginning users. The book covers both the hardware and software ends of the chip's operation. Once the basic theory is established, the majority of BASIC Stamp walks you through applications suitable for designers as well as the home hobbyist. These applications can be used as is, or as a basis for further modifications to suit your needs.

www.ingramcontent.com/pod-product-compliance
Lightning Source LLC
Chambersburg PA
CBHW030918180526
45163CB00002B/383